Energy Arbitration and Judicial Dispute Settlement

Veröffentlichungen des Instituts für Energie- und Regulierungsrecht Berlin
Herausgegeben von Franz Jürgen Säcker und Jochen Mohr

Band 72

Jochen Mohr / Antonis Metaxas (eds.)

Energy Arbitration and Judicial Dispute Settlement

Proceedings of the 4th Athens Conference on European Energy Law

Bibliografische Information der Deutschen Nationalbibliothek
Die Deutsche Nationalbibliothek verzeichnet diese Publikation
in der Deutschen Nationalbibliografie; detaillierte bibliografische
Daten sind im Internet über http://dnb.d-nb.de abrufbar.

Library of Congress Cataloging-in-Publication Data
A CIP catalog record for this book has been applied for at the
Library of Congress.

ISSN 2363-684X
ISBN 978-3-631-81588-5 (Print)
E-ISBN 978-3-631-83644-6 (E-PDF)
E-ISBN 978-3-631-83645-3 (EPUB)
E-ISBN 978-3-631-83646-0 (MOBI)
DOI 10.3726/b17632

© Peter Lang GmbH
Internationaler Verlag der Wissenschaften
Berlin 2021
Alle Rechte vorbehalten.

Peter Lang – Berlin · Bern · Bruxelles · New York ·
Oxford · Warszawa · Wien

All parts of this publication are protected by copyright. Any
utilisation outside the strict limits of the copyright law, without
the permission of the publisher, is forbidden and liable to
prosecution. This applies in particular to reproductions,
translations, microfilming, and storage and processing in
electronic retrieval systems.

This publication has been peer reviewed.

www.peterlang.com

Editors' preface

On September 27, 2019, the Institute for Energy und Regulatory Law Berlin (enreg) and the Hellenic Energy Regulation Institute hosted in Athens an international congress on Energy Arbitration and Judicial Dispute Settlement. Both editors had the pleasure of acting as the scientific coordinators/directors of this important event that gathered in Athens some of the most leading scientists in the field of Energy Regulation and Arbitration in Europe.

The contributions in this book reflect the remarkable width and depth of the subjects dealt with, spanning from the Representative Actions in the energy sector over the landmark *Achmea* judgment of the CJEU and its far reaching impact on Energy Arbitration to the dynamic relationship between the notion of autonomy of EU Law and judicial dispute settlement in the field of energy-related investments. Our hope is that this publication substantially contributes to the scientific discussion about the future design of arbitration in the EU and the prospects of legal protection of international investments in the European energy sector.

The editors cordially thank all the leading senior judges, practitioners and academics that contributed to this collective volume as well as the members of the Hellenic Energy Regulation Institute and of the Institute for Energy und Regulatory Law Berlin for their active support in preparing this publication, above all Dr. Konstantina Bourazeri, Michalis Mathioulakis, Till Meier and Kim Natalie Müller.

Berlin / Athens, May 2020
Prof. Dr. Jochen Mohr / Prof. Dr. Antonis Metaxas

Contents

Prof. Dr. Jochen Mohr & Dr. Konstantina Bourazeri, LL.M.
1 The impact of *Achmea* and CETA on investment arbitration 15

    Introduction 15

    1 The *Achmea* ruling of the ECJ 19
        1.1 Subject 19
        1.2 Key statements 20
        1.3 Legal consequences 22

    2 Opinion 1/17 of the ECJ on the CETA Agreement 24
        2.1 Subject 24
        2.2 Key statements 24
        2.3 Legal consequences 25

    3 Investment protection under the Energy Charter Treaty and ECJ jurisprudence in *Achmea* and CETA 26
        3.1 Overview of the Energy Charter Treaty 26
        3.2 Effects of *Achmea* and CETA jurisprudence on investment arbitration based on the Energy Charter Treaty 27

    Conclusion in Theses 30

Prof. Dr. Antonis Metaxas
2 Safeguarding the autonomy of EU Law: The impact of ISDS mechanisms after Opinion 1/17 31

    Introduction 31

    1. The concept of autonomy 33

    2. The fundamental aspects of the principle of autonomy in the light of Opinion 1/17 35
        2.1 Allocation of competences 35
        2.2 The significance of the preliminary reference procedure ... 37
        2.3 Control over the uniform application of EU law 42

    3. Evaluating the principle of autonomy after Opinion 1/17 44

    Conclusions 46

Wiegand Laubenstein
3 Judicial settlement of disputes in the energy sector in Germany 49

   1 The history of German regulatory law 49

   2 Higher Regional Court of Düsseldorf 52

   3 The *Lichtblick* case 53

   4 Incentive regulation of the network operators – the X-factor 54

      4.1 Relevant observation period (Stützintervall) 57

      4.2 Analogue application of the provisions about individual efficiency . 57

   5 Poland wins OPAL gas pipeline dispute before EU court 58

      5.1 The facts 58

      5.2 Further procedure 61

Theodoros Galanis
4 Multilateral Investment Court – an alternative to energy disputes 63

   Introduction 63

   1 Main criticism addressed to ISDS 65

   2 The new multilateral approach 67

   3 The Multilateral Investment Court 72

   4 Basic features of MIC – the Council mandate 73

   5 Critical issues – some reflections 74

      5.1 Critical issues – some answers 76

   Concluding remarks 79

Agis Chr. Georgiades
5 Energy dispute settlement after *Achmea* and opinion 1/17 81

   Introduction 81

   1 Dispute settlement in bilateral and multi-lateral investment treaties 82

   2 Energy Charter Treaty 83

   3 The *Achmea* case 83

      3.1 The arguments of Slovakia and the European Commission 84

      3.2 CJEU on *Achmea* 85

4 Opinion 1/17 of the CJEU on CETA	87
5 The prevailing interpretation of *Achmea*	88
6 Criticism against *Achmea*	90
7 Adequacy of EU response	92
Conclusions	92

Louise Bouvery
6 The use of technically based alternative dispute resolution mechanisms in oil & gas contracts 95

Introduction 95

   1 The new methods: Technical Advisory Committees, Dispute Review Boards, Sole Expert Determination 97

      1.1 General characters of the mechanisms 97

         1.1.1 Technical Advisory Committees 97

         1.1.2 DRBs 99

         1.1.3 Sole Expert Determination 100

      1.2 Dispute prevention 103

   2 Interaction between arbitration and the unconventional ADRMs 104

      2.1 Unconventional ADRMs as an alternative to arbitration 104

      2.2 Unconventional ADRMs as a prior step to arbitration 107

   3 Considering a multi-tiered dispute resolution clause 109

Conclusions 111

Dr. Anna Plevri
7 Dispute settlement resolution mechanism in the Greek current exploration licensing rounds 113

   1 Disputes arising out of the exploration and exploitation of hydrocarbons and arbitration 113

   2 Latest developments in the Greek framework of exploration licensing rounds 118

  3 Legislative framework of exploration and exploitation of hydrocarbons in Greece 120

    3.1 Law 2289/1995 (Hydrocarbons Law) as amended by Law 4001/2011 120

    3.2 Presidential decree No 127/1996 and Draft Model Lease Agreement 123

  4 Calls for international tenders for the right of exploration and exploitation of hydrocarbons in areas of Greece 125

  5 Laws 4524/2018, 4525/2018 4526/2018 and 4527/2018 on the ratification of lease agreements between the Greek State and various energy companies on the grant of the right to explore and exploit hydrocarbons in various lands or marine areas in Greece 127

    5.1 Resolution of disputes 128

  6 Variations of Law 4525/2018 on issues of the dispute settlement resolution mechanism 130

  Conclusions 132

Michalis Mathioulakis
8 The role of legal disputes and arbitration in EU's energy security: The case of natural gas networks 135

  Introduction 135

  1 The role of legal disputes and arbitration in natural gas imports for the European energy security 136

  2 Disputes and arbitration over gas imports in Europe 137

    2.1 The OPAL case 137

    2.2 Gazprom – PGNiG arbitration 139

    2.3 Gazprom – European Commission notice of dispute 140

    2.4 Gazprom – Naftogaz arbitration 142

  Conclusions 145

Sofia Michelaki
9 Claims of users concerning capacity charges of Natural Gas Transmission System. Case law 147

  Introduction 147

  1 Recent arbitration cases regarding capacity charges 148

2 Legal analysis 149
    2.1 Legislative Framework 149
    2.2 Factual background with regard to the case 150
  Conclusions 153

Peter Mussaeus & Adriane Mönch
10 Germany's electricity price zones and their delineations 155
  Introduction 155
  1 The General Court's decision on capacity bookings 156
  2 Phasing out of coal-fired power plants 157
  3 The Energy Charter Treaty 158
  4 Electricity price zones 159
    4.1 The problem of loop-flows 160
    4.2 The new Article 14 Regulation (EU) 2019/943 163
  5 A conflict with § 3a StromNZV? 165
  Conclusions 167

Alexia Trokoudi & Emmanouil Chourmousis
11 The General Court's Decision of 15 November 2018 (case
  T-793/14) for the annulment of the EC decision of 23 July 2014
  (case S. 35980) on the Capacity Mechanism in the UK 169
  Introductory note 170
  1 Commission's Decision C (2014) 5083 approving the UK
    Capacity Mechanism 170
    1.1 The mechanism 170
    1.2 Commission's assessment 172
    1.3 The judgment of the EU General Court 172
  2 Procedural steps before the Court – Admissibility of the action 173
  3 Court's legal considerations on the substance of the case 173
    3.1 Concept of doubts 173
    3.2 Length of discussions between the Commission and the UK 174

3.3 Assessment of the DSR role in the UK capacity market 175
3.4 Discriminatory treatment of DSR in the UK capacity market 175

4 The procedural events following the annulment of the UK capacity mechanism 177

Conclusions 178

Dr. Maria Lykidi
12 Arbitration in the construction of nuclear power plants 179

1 Main issues resulting in arbitration in the nuclear sector 179

2 Construction arbitration in the nuclear sector – The case of EPR 182

 2.1 Olkiluoto (OL3) – Finland 185

 2.2 Flamanville – France 186

 2.3 Hinkley Point – United Kingdom 187

3 European Court of Justice's decision on environmental impact of lifetime prolongation of nuclear power plants – The case of Belgium and the Netherlands 191

Alexandros Chatzinerantzis
13 Climate-change related disputes – Overview and trends 195

Introduction 195

Range of climate-related disputes 196

1 Climate enforcement disputes 196

 1.1 Cases brought forward against government agencies to change climate policy and conduct 197

 1.2 Cases brought forward to seek damages for financial redress 199

2 Commercial litigation and arbitration 201

3 Investor-State arbitration 202

4 Investor and D&O claims 204

Conclusions 206

Prof. Dr. Matthias Knauff, LL.M. Eur.
14 Representative actions in the energy sector 207

    Introductory remarks 207

    1 Representative actions regarding environmental matters 208
        1.1 Development 208
        1.2 Design 209
        1.3 The problem of preclusion 211

    2 Proposal for a directive on representative actions for the protection of the collective interests of consumers 214
        2.1 Scope 216
        2.2 Design 217
        2.3 Consequences for the energy sector 221

    Conclusions 222

Dr. Athina Moraiti
15 Consumer protection and the right to out-of-court dispute settlement in the oil & gas sector. Regulator or Ombudsman model for Energy Alternative Dispute Resolution in Europe? 223

    1 Consumer rights and protection 223

    2 Oil & gas industry disputes 229

    3 Types of disputes – under examination only disputes between individuals and oil & gas companies 230

    4 Methods of dispute resolution 230

    5 Ombudsman – mediation 232

    6 Regulators' special competences to resolve disputes 233

    7 Which is the best model for Europe to adopt in oil & gas disputes? 237

Dr. Kyriakos P. Papanikolaou
16 Energy investments under changed circumstances: Emerging gaps and necessities in arbitration 239

    1 Long-term energy contracts under a change of circumstances 239

    2 The distinction between *rebus sic stantibus* and *force majeure* 240

    3 *Rebus sic stantibus* vs. *Pacta sunt servanda* 241

    4 The reluctance of arbitral tribunals to apply the principle *rebus sic stantibus* 243

Prof. Dr. Jochen Mohr & Dr. Konstantina Bourazeri, LL.M.*

1 The impact of *Achmea* and CETA on investment arbitration

Abstract: This chapter discusses the *Achmea* ruling of the European Court of Justice (ECJ) from 2018 concerning bilateral investment protection agreements between the EU Member States and its opinion on CETA from 2019. It focuses on the dogmatic importance of ECJ's recent jurisprudence and outlines its possible effects on investment arbitration under the Energy Charter Treaty.

Keywords: *Achmea*, CETA, intra-EU BITs, international investment arbitration, bilateral investment protection agreements (BITs), Energy Charter Treaty

Introduction

The compatibility of investment arbitration under international law with Union law has been subject to a lively debate for years now. The discussion has recently been intensified by the ECJ's decision on *Achmea*[1] and its opinion on

---

* Jochen Mohr is Holder of the Chair of Civil Law, Competition Law, Energy Law, Regulatory Law and Labour Law at the University of Leipzig and director of the Institut für Energie- und Regulierungsrecht Berlin e. V. (enreg). Konstantina Bourazeri is Attorney at Law and Lecturer at the University of Leipzig.

1 European Court of Justice (Grand Chamber), judgment of 6 March 2018, case C-284/16, ECLI:EU:C:2018:158, *Achmea*. On this decision see Jörg Gundel, "Investitionsschutz-Schiedsgerichtsbarkeit und Unionsrecht nach dem *Achmea*-Urteil des EuGH*", Europäisches Wirtschafts- und Steuerrecht* (2018), 124; Claus Dieter Classen, "Autonomie des Unionsrechts als Festungsring? – Anmerkung zum Urteil des EuGH (GK) v. 6.3.2018, Rs. C-284/16 (Slowakische Republik/ Achmea BV)", *Europarecht* (2018), 361; Robin Dominik Miller, "Autonomie des Unionsrechts versus Schiedsgerichtsbarkeit", *Europäische Zeitschrift für Wirtschaftsrecht* (2018), 357; Patricia Sarah Stöbener de Mora, "Das Achmea-Urteil zum Intra-EU-Investitionsschutz – Die Auswirkungen und die Notwendigkeit eines EU-weiten Schutzmechanismus", *Europäische Zeitschrift für Wirtschaftsrecht* (2018), 363; Jens Brauneck, "Multilateraler Gerichtshof und EuGH-Achmea-Urteil: Das Ende aller EU-mitgliedstaatlich vereinbarten Schiedsgerichte", *Europarecht* (2018), 429; Joscha Müller and Sven Simon, "Das Achmea-Urteil des EuGH und die Auswirkungen auf Streitbeilegungsmechanismen im Rechtsraum der EU", *Neue Juristische Online-Zeitschrift* (2018), 961; Stephan Wernicke,

the Comprehensive Economic and Trade Agreement, shortly CETA.[2] In the *Achmea* judgment from 2018, the ECJ dealt with bilateral investment protection agreements between the EU Member States, the so-called intra-EU BITs.[3] In the past, these agreements have been concluded between the individual EU Member States and non-member countries of Central and Eastern Europe.[4] In the time of the *Achmea* decision, there were almost 200 intra-EU BITs in effect,[5] the oldest one between Germany and Greece dating back to 1961.[6] The BITs protect foreign investors against discriminatory treatment or expropriation

---

"Autonomie und Häresie – Investitionsschiedsgerichte in der Rechtsunion", *Neue Juristische Wochenschrift* (2018), 1644; Nikos Lavranos and Tania Singla, "Achmea: Groundbreaking or Overrated?", *Zeitschrift für Schiedsverfahren* (2018), 348; Andrej Lang, "Die Autonomie des Unionsrechts und die Zukunft der Investor-Staat-Streitbeilegung in Europa nach Achmea. Zugleich ein Beitrag zur Dogmatik des Art. 351 AEUV", *Europarecht* (2018), 525; Claas Friedrich Germelmann, "Die Zukunft der internationalen Schiedsgerichtsbarkeit im Lichte der jüngsten EuGH Rechtsprechung", in: *Grundfragen des (Energie-)Regulierungsrechts und des Wirtschaftsrechts – Festveranstaltung aus Anlass des 80. Geburtstags von Prof. Dr. Gunther Kühne*, eds. Jochen Mohr and Harmut Weyer (expected date of publication summer 2020, Peter Lang), pp. 95–97, 99 et seq.

2 European Court of Justice, opinion 1/17 of 30 April 2019, ECLI:EU:C:2019:341, *CETA*; see on this opinion Jörg Gundel, "Das CETA-Gutachten des EuGH: Neue Grenzen des Unionsrechts für die Unterwerfung unter 'fremde Richter'?", *Europäisches Wirtschafts- und Steuerrecht* (2019), 181; Matthias Ruffert, "Zulässigkeit der CETA-Investitionsstreitbeilegung – Das CETA-Gericht und seine Rechtsbehelfsinstanz sind unionsrechtskonform", *Juristische Schulung* (2019), 598; Claas Friedrich Germelmann, in: *Grundfragen des (Energie-)Regulierungsrechts und des Wirtschaftsrechts*, pp. 97–99, 112–113; Daniel Engel, "Vereinbarkeit unionsexterner Investitionsschutzabkommen mit der Autonomie der Unionsrechtsordnung – der (vorerst) letzte Baustein der EuGH-Rechtsprechung zum EU-Investitionsschutz", *Zeitschrift für Schiedsverfahren* (2019), 231 et seq.; Boris Scholtka, "Anmerkung zu EuGH (Plenum), Gutachten v. 30.4.2019 – Avis 1/17", *Neue Zeitschrift für Verwaltungsrecht* (2019), 868; especially concerning the possibility of arbitration courts to refer a question to the ECJ under the preliminary ruling procedure cf. Jörg Gundel, "Die Öffnung des Vorabentscheidungsverfahrens zum EuGH für nichtmitgliedstaatliche Gerichte", *Europäische Zeitschrift für Wirtschaftsrecht* (2019), 934, 935 et seq., 938.

3 Bilateral Investment Treaty – BIT.

4 Gundel, *Europäisches Wirtschafts- und Steuerrecht* (2018), 124 and 125.

5 Opinion of AG Wathalet delivered on 19 September 2017, case C-284/16, ECLI:EU:C:2017:699, *Achmea*, para. 3.

6 Opinion of AG Wathalet delivered on 19 September 2017, case C-284/16, ECLI:EU:C:2017:699, *Achmea*, para. 3 with fn. 6.

without compensation by the "host State", in which the investment takes place.[7] In this context, it is particularly important that the BITs often provide for arbitration clauses which allow private investors from one Contracting State to settle disputes with the other Contracting State before an ad hoc arbitral tribunal.[8] This avoids the way to the ordinary courts of the host State. Due to the later accession of the non-European Contracting Parties to the EU, particularly in the course of the EU enlargement in 2004, the regular BITs became intra-EU BITs. The European Commission has been long supporting the view that intra-EU BITs are incompatible with Union law,[9] advocating the establishment of a permanent multilateral court of arbitration at Union level instead.[10] With its *Achmea* ruling of 2018, the ECJ has strengthened the Commission's perspective. The ECJ does not generally consider intra-EU BITs to be contrary to European law. However, it places such high demands on their compatibility that this de facto leads to their general inapplicability. In the aftermath of the *Achmea* ruling, 22 member states – including Germany and Greece – have therefore agreed to terminate their existing BITs on a bilateral or even multilateral basis by the end of 2019.[11] Following this declaration, an ad hoc group of EU Member States coordinated by the European Commission has developed

---

7   Boris Scholtka, "Comment on ECJ's judgment *Achmea*", *Europäische Zeitschrift für Wirtschaftsrecht* (2018), 243; the importance of BITs for medium-sized businesses emphasises Stephan Wernicke, *Neue Juristische Wochenschrift* (2018), 1644, 1645.
8   Scholtka, *Europäische Zeitschrift für Wirtschaftsrecht* (2018), 243; in particular on the relationship between investment arbitration and diplomatic protection see Jörg Gundel, "Vom diplomatischen Schutz zum Recht der Investitionsschutzabkommen: Offene Fragen und alternative Wege beim Auslandsschutz privater Vermögensinteressen", *Archiv des Völkerrechts* 51 (2013), 108, 133 et seq.
9   Concerning the view of the European Commission, see opinion of AG Wathalet delivered on 19 September 2017, case C-284/16, ECLI:EU:C:2017:699, *Achmea*, para. 3 and 39 et seq.; Gundel, *Europäisches Wirtschafts- und Steuerrecht* (2018), 124, 125.
10 European Commission, Recommendation for a Council Decision authorising the opening of negotiations for a Convention establishing a multilateral court for the settlement of investment disputes, COM(2017) 493 final of 13 September 2017; European Commission, The Multilateral Investment Court project, http://trade.ec.europa.eu/doclib/press/index.cfm?id=1608 (accessed 15 April 2020).
11 "Declaration of the representatives of the Governments of the Member States of 15 January 2019 on the legal consequences of the judgment of the Court of Justice in Achmea and on investment protection in the European Union", https://ec.europa.eu/info/sites/info/files/business_economy_euro/banking_and_finance/documents/190117-bilateral-investment-treaties_en.pdf (accessed 15 April 2020).

a "[Draft-]Agreement for the Termination of Bilateral Investment Treaties between the Member States of the European Union.[12] This plurilateral treaty stipulates the termination of all ca. 190 intra-EU BITs, including their sunset clauses.[13] Against this background, the question arises, which is the impact of *Achmea* on another important category of investment arbitral tribunals between Member States, the ones based on the multilateral Energy Charter Treaty of 1994.[14] Intra-EU BITs should be distinguished from multilateral investment protection agreements, which provide for permanent dispute settlement mechanisms instead of ad hoc arbitration. Such an agreement was subject of the ECJ's opinion 1/17 from 2019, in which the ECJ examined whether the Comprehensive Economic and Trade Agreement concluded between Canada, the EU and its Member States was compatible with European law. In contrast to the *Achmea* case, the ECJ did not find any impairment of the autonomy of the Union legal system, mainly due to the institutional structure of the CETA-courts and their limited powers of control. In the following, we will first present the dogmatic core statements of the ECJ's *Achmea* ruling (II.). Then we will outline the opinion 1/17 of the Court on the CETA agreement (III.). Subsequently, we will discuss possible effects of the ECJ's judicature on the intra-EU investment protection procedures under the Energy Charter Treaty (IV.).

---

12 This draft agreement has not been officially published yet. The text of the agreement is available in: *Zeitschrift für Schiedsverfahren* (2020), 85; an online version including the full text of the Annexes is available at: http://arbitrationblog.kluwerarbitration.com/wp-content/uploads/sites/48/2019/12/a-draft-agreement-has-been-leaked.pdf (accessed 15 April 2020).
13 Annexes A and B of the agreement; Editorial Note, *Zeitschrift für Schiedsverfahren* (2020), 85.
14 In relation to the rules of the Energy Charter Treaty for ad hoc arbitral tribunals concerning the settlement of disputes between a Contracting Party and an Investor of another Contracting Party (Article 26) and disputes between Contracting Parties (Article 27) see Gundel, in: *Berliner Kommentar zum Energierecht*, ed. Franz Jürgen Säcker (Frankfurt am Main: Fachmedien Recht und Wirtschaft, 2019), Vol. 1, Einleitung D EnWG, para. 35 et seq.; regarding the impact of *Achmea* on investor-state dispute settlement on the basis of the Energy Charter Treaty see further Claas Friedrich Germelmann, in: *Grundfragen des (Energie-)Regulierungsrechts und des Wirtschaftsrechts*, pp. 108 et seq.

1 The *Achmea* ruling of the ECJ

1.1 Subject

The subject of the *Achmea* ruling of March 2018[15] was a bilateral investment protection agreement between the Netherlands and the Slovak Republic. Article 8 of this BIT provided for the possibility of disputes between one Party and an investor of the other Party before an arbitral tribunal. On the basis of this BIT, *Achmea*, a Dutch insurance group, initiated arbitration proceedings against the Slovak Republic in October 2008.[16] In these proceedings, *Achmea* claimed compensation for damage suffered by one of its subsidiaries – a provider of private health insurance in Slovakia – as a result of measures taken by the Slovak Republic.[17] These state measures aimed at partially reversing the liberalisation of the private health insurance market. The arbitration proceedings ended in December 2012 with a positive verdict for *Achmea*, which ordered the Slovak Republic to pay damages to the aforementioned subsidiary company of *Achmea*.[18] Since the arbitral tribunal had chosen the German city of Frankfurt am Main as the place of arbitration, and the proceedings were, therefore, subject to German law, the Slovak Republic challenged the arbitral award by means of German law. So, the Slovak Republic applied to the Higher Regional Court of Frankfurt am Main requiring the annulment of the arbitral award.[19] After the rejection of that application, the Slovak Republic appealed

---

15 European Court of Justice (Grand Chamber), judgment of 6 March 2018, case C-284/16, ECLI:EU:C:2018:158, *Achmea*.
16 See opinion of AG Wathalet delivered on 19 September 2017, case C-284/16, ECLI:EU:C:2017:699, *Achmea*, para. 17 et seq.
17 See in detail Arbitral tribunal, award of 7 December 2012, PCA Case No. 2008–13, *Achmea B.V. (formerly Eureko B.V.) v. The Slovak Republic (I)*, para. 91 et seq., http://www.italaw.com/sites/default/files/case-documents/italaw3206.pdf (accessed 15 April 2020).
18 Arbitral tribunal, award of 7 December 2012, PCA Case No. 2008–13, *Achmea B.V. (formerly Eureko B.V.) v. The Slovak Republic (I)*, para. 331 et seq.
19 Higher Regional Court of Frankfurt am Main, decision of 18 December 2014, case 26 Sch 3/13; with respect to Slovak Republic's application for annulment against the positive interim decision of the Arbitration Court of 26 October 2010 on its jurisdiction, see Higher Regional Court of Frankfurt am Main, decision of 10 May 2012, case 26 SchH 11/10, *Zeitschrift für Schiedsverfahren* (2013), 119; concerning this decision see further Jan K. Schäfer and John P. Gaffney, "Intra-EU BITs: Toothless Tigers or Do They Still Bite? The OLG Frankfurt Considers the Impact of EU Law on the Investor-State Dispute Resolution Mechanism", *Zeitschrift für Schiedsverfahren* (2013), 68 et seq.

to the German Federal Court of Justice, the "Bundesgerichtshof".[20] The Slovak Republic claimed that the arbitral award as well as the arbitration agreement on which it was based, were contrary to the public order.[21] In particular, according to the Slovak Republic, the arbitration procedure was incompatible with the Articles 267 and 344 TFEU, which ensure the uniform application of Union law.[22] The German Federal Court of Justice did not consent with the complaints of the Slovak Republic and thus rejected the appeal.[23] However, due to the great importance of the legal questions arising in this matter, it requested a preliminary ruling by the ECJ.[24]

1.2 Key statements

Unlike the German Federal Court of Justice and Advocate General *Wathalet*,[25] the ECJ found that the arbitration clause of the BIT at issue was incompatible with EU law. Articles 267 and 344 TFEU preclude a provision, such as Article 8 of the BIT, under which an investor from a Member State may initiate proceedings before an arbitral tribunal in disputes concerning investments in another Member State.[26] To justify this finding, the ECJ relies on the autonomy of the

---

20 Bundesgerichtshof, decision of 3 March 2016, case I ZB 2/15, *Zeitschrift für Schiedsverfahren* (2016), 328, para. 8.
21 See opinion of AG Wathalet delivered on 19 September 2017, case C-284/16, ECLI:EU:C:2017:699, *Achmea*, para. 26 et seq.
22 For details of the Slovak Republic's legal argumentation see Higher Regional Court of Frankfurt am Main, decision of 18 December 2014, case 26 Sch 3/13, para. 29 and 30; Higher Regional Court of Frankfurt am Main, decision of 10 May 2012, case 26 SchH 11/10, *Zeitschrift für Schiedsverfahren* (2013), 119, 120; for an overview of Slovak Republic's appeals before the German courts see Bundesgerichtshof, decision of 3 March 2016, case I ZB 2/15, *Zeitschrift für Schiedsverfahren* (2016), 328, para. 7 and 8.
23 Bundesgerichtshof, decision of 3 March 2016, case I ZB 2/15, *Zeitschrift für Schiedsverfahren* (2016), 328, para. 24, 45 and 78; see also Gebhard Bücheler and Igor Vasiljević, "Bitte kein BIT? Investitionsschutzverträge auf dem Prüfstand des Unionsrechts", *Zeitschrift für Schiedsverfahren* (2016), 336.
24 Bundesgerichtshof, decision of 3 March 2016, case I ZB 2/15, *Zeitschrift für Schiedsverfahren* (2016), 328, para. 19 et seq.
25 Opinion of AG Wathalet, delivered on 19 September 2017, case C-284/16, ECLI:EU:C:2017:699, *Achmea*, para. 54 et seq.
26 European Court of Justice (Grand Chamber), decision of 6 March 2018, case C-284/16, ECLI:EU:C:2018:158, *Achmea*, para. 31 et seq., 60. The Bundesgerichtshof [decision of 31 October 2018, case I ZB 2/17, *Zeitschrift für Schiedsverfahren* (2019), 46] followed the ECJ's judgment and set aside the *Achmea* arbitration award; see Stephan Wilske

Union legal system. According to Article 344 TFEU, Member States undertake not to submit a dispute concerning the interpretation or application of the Treaties to any method of settlement other than those provided for in the European Treaties. International agreements should, therefore, not affect the system of jurisdiction established by Union law.[27] In order to underpin the autonomy of Union law in relation to the Member States and international law, the ECJ highlights the constitutional structure of the Union and the nature of Union law as an autonomous source of law.[28] The Court emphasizes that the EU judicial system laid down in Article 19 of the TEU ensures a uniform interpretation of Union law. 'Key element' of this judicial system is the preliminary ruling procedure according to Article 267 TFEU.[29] The extent to which intra-EU BITs are compatible with the Union legal system depends on whether an arbitral tribunal has to deal with the application or even interpretation of Union law.[30] The Court attaches central importance to Article 8 Section 6 of the BIT, according to which the arbitral tribunal must take into account the applicable law of the Contracting Party concerned and all relevant agreements between the Contracting Parties. According to the ECJ, this provision gives the arbitral tribunal the competence both to interpret and apply Union law, since Union law is an integral part of the Member State's law.[31] However, such

---

and Lars Markert, in: *Beck'scher Online Kommentar Zivilprozessordnung*, eds. Volkert Vorwerk and Christian Wolf (München: C.H. Beck, 2020), § 1059 ZPO, para. 36.1.

27 European Court of Justice (Grand Chamber), decision of 6 March 2018, case C-284/16, ECLI:EU:C:2018:158, *Achmea*, para. 32, referring to Court of Justice, opinion 2/13 of 18 December 2014, EU:C:2014:2454, *Accession of the European Union to the European Convention for the Protection of Human Rights and Fundamental Freedoms*, para. 201; regarding the legal consequences of a potential accession of the EU to the ECHR for the European fundamental rights protection cf. Konstantina Bourazeri, "Tarifautonomie und Wirtschaftskrise" (Baden-Baden: Nomos, 2019), pp. 281 et seq. In its opinion 2/13 (op. cit.), the ECJ considered, as in its opinion 1/09 of 8 March 2011 (ECLI:EU:C:2011:123) regarding the creation of a European Patent Court, the interest of safeguarding the autonomy of Union law as prior to the interest of integrating international law dispute settlement mechanisms in the Union legal order. The *Achmea* judgment follows this view; see Andrej Lang, *Europarecht* (2018), 525, 528, 531 et seq.

28 European Court of Justice (Grand Chamber), decision of 6 March 2018, case C-284/16, ECLI:EU:C:2018:158, *Achmea*, para. 33.

29 Ibid., para. 35 et seq.

30 Ibid., para. 39 and 40.

31 Ibid., para. 41 et seq.

competence can only be compatible with the EU legal system if the arbitral tribunal is a 'court of a Member State' in terms of Article 267 TFEU. The ECJ denies that intra-EU arbitral tribunals hold this status.[32] Consequently, arbitral tribunals cannot make a reference to the ECJ for a preliminary ruling, even though they decide on disputes which might require the application of Union law.[33] This is a core point in the ECJ's argumentation. An arbitral award can, however, be under judicial control through the Member States courts which – in contrast to arbitral tribunals – are able to refer questions to the ECJ for a preliminary ruling and thus contribute to the uniform interpretation of EU law.[34] Nevertheless, the German law applicable to the *Achmea* case only allows a limited substantive-law review of arbitral awards, especially if their recognition and enforcement comply with the European and German public order (§ 1059 para 2 no. 2 lit. b German Code of Civil Procedure).[35] In the ECJ's view, this limited control is not suitable for ensuring the full effectiveness of Union law.[36] Last but not least, the ECJ considers the submitting of investor-state-disputes to ad hoc arbitral tribunals, which are not part of the EU judicial system, a violation of the principle of loyal cooperation between the Member States (Article 4 para. 3 subpara. 1 TEU).[37] As we will see later on, this argument is significant for the legal assessment of intra-EU disputes on the basis of the Energy Charter Treaty.

1.3 Legal consequences

According to the *Achmea* judgment, investor-state arbitration based on intra-EU BITs is contrary to Union law insofar as the arbitral tribunals have to deal with the interpretation of Union law. This conflicts with the lack of authority of the arbitral tribunals to appeal to the ECJ for a preliminary ruling. The ECJ focuses on the abstract threat of its primacy of interpretation, regardless of whether EU law is actually relevant in a specific case or not.[38]

---

32 This is criticised by Jens Brauneck, *Europarecht* (2018), 429, 449 et seq.
33 European Court of Justice (Grand Chamber), decision of 6 March 2018, case C-284/16, ECLI:EU:C:2018:158, *Achmea*, para. 45 et seq., 49.
34 Ibid., para. 50 et seq.
35 Ibid., para. 53.
36 Ibid., para. 54 et seq.
37 Ibid., para. 58.
38 This is also emphasised by Robin Dominik Miller, *Europäische Zeitschrift für Wirtschaftsrecht* (2018), 357, 359; see further the critical thoughts of Patricia Sarah Stöbener de Mora, *Europäische Zeitschrift für Wirtschaftsrecht* (2018), 363, 364.

The *Achmea* judgment is therefore relevant for all intra-EU BITs that provide for arbitration.[39] Another striking argument of the ECJ is the principle of loyal cooperation; in other words, the principle of loyalty to the EU. This principle requires mutual trust between the Member States in the courts of other Member States as core components of the Union legal system. This mutual trust, though, is undermined by the establishment of arbitral tribunals, according to the ECJ.[40] From this point of view, it becomes apparent, why the ECJ rejects investor-state arbitration between the Member States, whereas it recognises – in an obiter dictum – commercial arbitration based on private autonomy.[41]

The *Achmea* judgment raises many questions. An important one is the ECJ's formal focus on the principle of loyalty to the EU. The accession of the Central and Eastern European countries to the EU does not mean that investment protection by means of arbitral tribunals is nowadays no longer justified. The possibility to settle disputes between an investor and the host State before a neutral body still remains an important function of investment arbitration.[42] From a foreign investor's point of view, who is raising financial claims against the host State, an arbitral tribunal can be more suitable than the ordinary courts of the host State, as the latter may be "institutionally biased".[43] Another issue raised by the *Achmea* ruling is that the ECJ has not limited the legal effects of its decision in terms of time. The assessment that a case is contrary to Union law, therefore, not only affects ongoing arbitration proceedings,[44] but can also lead to a rescission of proceedings that have already been closed.[45] Such a retrospective effect can have more intensive consequences than a contract termination, as the latter does not influence the protection of investments guaranteed

---

39 Müller and Simon, *Neue Juristische Online-Zeitschrift* (2018), 961, 963; Lavranos and Singla, *Zeitschrift für Schiedsverfahren* (2018), 348, 350.
40 Cf. European Court of Justice (Grand Chamber), decision of 6 March 2018, case C-284/16, ECLI:EU:C:2018:158, *Achmea*, para. 34, 58; European Court of Justice, opinion 1/17 of 30 April 2019, ECLI:EU:C:2019:341, *CETA*, para. 128.
41 European Court of Justice (Grand Chamber), decision of 6 March 2018, case C-284/16, ECLI:EU:C:2018:158, *Achmea*, para. 55; see also Gundel, *Europäisches Wirtschafts- und Steuerrecht* (2018), 124, 126.
42 Gundel, *Europäisches Wirtschafts- und Steuerrecht* (2018), 124, 127.
43 Claus Dieter Classen, "Die Unterwerfung demokratischer Hoheitsgewalt unter eine Schiedsgerichtsbarkeit", *Europäische Zeitschrift für Wirtschaftsrecht* (2014), 611, 613.
44 Brauneck, *Europarecht* (2018), 429, 431 and 432.
45 Gundel, *Europäisches Wirtschafts- und Steuerrecht* (2018), 124, 126 and 131.

through contractual "sunset clauses".[46] However, the aforementioned "[Draft-] Agreement for the termination of Bilateral Investment Treaties between the Member States of the European Union" stipulates in Article 2 para. 2 and Article 3 that sunset clauses of intra-EU BITs are terminated and shall not produce legal effects. Furthermore, the *Achmea* decision raised speculations on how the future of international investment arbitration will look like. In this context, the CETA opinion 1/17 of April 2019 gained special importance.

2 Opinion 1/17 of the ECJ on the CETA Agreement

2.1 Subject

In contrast to *Achmea*, the ECJ arrived at a positive assessment in its opinion 1/17 of April 2019[47] concerning the Comprehensive Economic and Trade Agreement (CETA) between Canada, the EU and its Member States that was signed in 2016.[48] CETA aims at a mutual opening of the markets of the Contracting Parties. It includes provisions on, inter alia, the protection of investments and the safeguarding of free and undistorted competition in the trading relations between the Parties.

2.2 Key statements

In opinion 1/17, the ECJ confirmed that the Investor-State Dispute Settlement system provided for in the CETA in relation to Canada is compatible with EU law. In his assessment, the ECJ points out the competence of the EU to submit itself, by concluding international agreements, to the decisions of courts entrusted with the interpretation of these agreements.[49] However, the binding

---

46 Gundel, *Europäisches Wirtschafts- und Steuerrecht* (2018), 124, 126. The "[Draft-] Agreement for the termination of Bilateral Investment Treaties between the Member States of the European Union" defines in Article 1 para. 7 as "sunset clause" any provision in a Bilateral Investment Treaty which extends the protection of investments made prior to the date of termination of that Treaty for a further period of time.
47 European Court of Justice, opinion 1/17 of 30 April 2019, ECLI:EU:C:2019:341, *CETA*.
48 Comprehensive Economic and Trade Agreement (CETA) between Canada, of the one part, and the European Union and its Member States, of the other part, Official Journal of the EU (OJ) L 11, 14 January 2017, 23; see also Fabian Stancke, "TTIP, CETA und die Rolle des Wettbewerbsrechts in internationalen Freihandelsabkommen", *Europäische Zeitschrift für Wirtschaftsrecht* (2016), 567, 569.
49 European Court of Justice, opinion 1/17 of 30 April 2019, ECLI:EU:C:2019:341, *CETA*, para. 106.

of the EU to international obligations must not impair the autonomy of the EU legal system. In this respect, the ECJ stresses – as in *Achmea* – the importance of the judicial system of the EU. The core element of this system is the exclusive jurisdiction of the ECJ over the definitive interpretation of EU law.[50] Although the dispute settlement mechanism provided for in CETA is outside the EU judicial system,[51] this does not preclude the compatibility of CETA dispute settlement mechanism with the Union legal system:[52] First, the CETA Tribunals are not given jurisdiction to interpret or apply Union law beyond the interpretation and application of the CETA provisions. Second, CETA Tribunals are not empowered to examine the "level of protection of a public interest" guaranteed by Union law on the basis of the CETA rules and to oblige the Union to pay damages on that basis.[53] Third, the CETA courts are accessible and independent within the meaning of Article 47 of the EU Charter of Fundamental Rights.[54]

2.3 Legal consequences

In its CETA opinion, the ECJ recognised the fundamental compatibility of international investor-state dispute settlements with EU law. At the same time, the ECJ made clear that *Achmea* referred solely to the special constellation of intra-EU BITs.[55] *Achmea*, therefore, does not mean a general rejection of international arbitration. Since CETA is a mixed agreement,[56] signed by the Union itself, the principle of loyalty to the Union is – in contrast to *Achmea* – irrelevant in this case, as this principle applies only to and between EU Member States. Furthermore, the ECJ clarified the *Achmea* jurisprudence

---

50 Cf. European Court of Justice, opinion 1/17 of 30 April 2019, ECLI:EU:C:2019:341, *CETA*, para. 111.
51 European Court of Justice, opinion 1/17 of 30 April 2019, ECLI:EU:C:2019:341, *CETA*, para. 112, 113 and 119.
52 European Court of Justice, opinion 1/17 of 30 April 2019, ECLI:EU:C:2019:341, *CETA*, para. 120 et seq., 137 et seq.
53 European Court of Justice, opinion 1/17 of 30 April 2019, ECLI:EU:C:2019:341, *CETA*, para. 152 et seq.
54 European Court of Justice, opinion 1/17 of 30 April 2019, ECLI:EU:C:2019:341, *CETA*, para. 189 et seq.
55 European Court of Justice, opinion 1/17 of 30 April 2019, ECLI:EU:C:2019:341, *CETA*, para. 127 et seq.
56 See Christian Tietje, "The Applicability of the Energy Charter Treaty in ICSID Arbitration of EU Nationals vs. EU Member States", *Essays on Transnational Economic Law* (2008), Issue 78, pp. 7 et seq.

by adding a second criterion to the one of the autonomy of Union's legal order: The scope of control exercised by an international investment arbitral tribunal. This control must be limited at such an extent that it does not adversely affect the powers of Union legislator to regulate the pursuit of legitimate public interests. Such negative effects are possible if impending compensation obligations due to a breach of the CETA guarantees could in fact force the EU to adjust law standards such as environmental or consumer protection.[57] Although this argumentation is generally convincing, the requirements set by the ECJ for the conformity of international dispute settlement with EU law limit the scope for international activity of the Union to a considerable extent.[58]

3 Investment protection under the Energy Charter Treaty and ECJ jurisprudence in *Achmea* and CETA

3.1 Overview of the Energy Charter Treaty

The Energy Charter Treaty (ECT) is an international agreement between the EU and 50 countries, including the EU Member States with the exception of Italy.[59] The treaty was concluded in Lisbon in 1994[60] and entered into force in 1998.[61] The Energy Charter Treaty aims to promote economic activities in the energy sector and facilitate international energy trade.[62] As of today, 122 arbitration proceedings have been initiated on the basis of the Energy Charter Treaty.[63] These include the currently pending *Vattenfall* case against Germany for accelerating

---

57 Jörg Gundel, *Europäisches Wirtschafts- und Steuerrecht* (2019), 181, 185.
58 So also Gundel, *Europäisches Wirtschafts- und Steuerrecht* (2019), 181, 188.
59 Claas Christian Germelmann, in: *Energierecht*, eds. Wolfgang Danner and Christian Theobald (München: C.H. Beck, 2019), Chapter 12 para. 25, 28.
60 OJ L 380, 31 December 1994, 3.
61 Gundel, in: *Handbuch des EU-Wirtschaftsrechts*, eds. Manfred A. Dauses and Markus Ludwigs (München: C.H. Beck, 2019), Chapter M. Energierecht, para. 206.
62 Germelmann, in: *Energierecht*, Chapter 12 para. 52; Gundel, in: *Handbuch des EU-Wirtschaftsrechts*, Chapter M. Energierecht, para. 209, 212; particularly with regard to the energy investment protection garantueed by the Energy Charter Treaty see Antonis Metaxas, "Investment Protection under the Energy Charter Treaty", in: *Transformation of EU and Eastern Mediterranean Energy Networks: Legal, Regulatory and Political Challenges*, eds. Leigh Hancher and Antonis Metaxas (Claeys & Casteels, 2018), pp. 23 et seq.
63 See the list of arbitration procedures at https://energycharter.org/what-we-do/dispute-settlement/all-investment-dispute-settlement-cases/ (accessed 15 April 2020).

the nuclear phase-out,[64] as well as several cases against Spain, Italy and the Czech Republic for cutting subsidies for renewable energies as a result of the financial crisis.[65] The majority of procedures initiated on the basis of the Energy Charter Treaty are intra-EU arbitration procedures.[66] In the light of the *Achmea* decision, the European Commission considers all procedures between Member States, even if they are based on the Energy Charter Treaty, as incompatible with EU law.[67] But does *Achmea* really lead to such a restriction for intra-EU arbitration based on the Energy Charter Treaty?

3.2 Effects of *Achmea* and CETA jurisprudence on investment arbitration based on the Energy Charter Treaty

Unlike the intra-EU BITs in *Achmea*, the Energy Charter Treaty has been signed as a mixed agreement by the Union and its Member States as well.[68] It is true that both the Union and the Member States must respect the exclusivity of the Union

---

64 ICSID, registered on 31 May 2012, case no. ARB/12/12, *Vattenfall AB et al v. Germany*; see on this case Markus Ludwigs, "Der Atomausstieg und die Folgen: Fordert der Paradigmenwechsel in der Energiepolitik einen Paradigmenwechsel beim Eigentumsschutz?", *Neue Zeitschrift für Verwaltungsrecht* (2016), 1, 4 et seq.; Jörg Gundel, "Völkerrechtliche Rahmenbedingungen der Energiewende – Der Energiecharta-Vertrag und das Vattenfall-Verfahren vor dem ICSID-Schiedsgericht", *Zeitschrift für das gesamte Recht der Energiewirtschaft* (2016), 243 et seq.; Julien Berger, "Die Bundesrepublik Deutschland – Internationaler Investitionsschutz und das Vattenfall-Verfahren", *Europäische Zeitschrift für Wirtschaftsrecht* (2020), 229, 230.
65 Gundel, "Frischer Wind für den Investitionsschutz nach dem Energiecharta-Vertrag?", *Zeitschrift für das gesamte Recht der Energiewirtschaft* (2018), 1, 2.
66 Julia Sackmann, "Im Schatten von CETA und TTIP: Zur Verfahrenstransparenz in Intra-EU-Investitionsschiedsverfahren", *Zeitschrift für Schiedsverfahren* (2015), 15, 17.
67 European Commission, Protection of intra-EU investment, COM(2018) 547 final of 19 July 2018, p. 4; critical to this view Friedmann Kainer, "Europäische Kommission: Investitionsschutz und Investitionsschutzabkommen innerhalb der EU", *Europäische Zeitschrift für Wirtschaftsrecht* (2018), 659; also Germelmann (in: *Grundfragen des (Energie-)Regulierungsrechts und des Wirtschaftsrechts*, p. 109) criticises as not persuasive the Commission's view, according to which the participation of the EU in the Energy Charter only refers to non-member countries; similar Tim Maxian Rusche, "Der Energiecharter-Vertrag auf dem Prüfstein von Achmea und CETA", *Europäische Zeitschrift für Wirtschaftsrecht* (2020), 169, 170.
68 The ratification of the ECT by the EU means, that in case the measure, which allegedly infringed investor rights, fell in the exclusive competence of the EU, the EU would have to act as respondent to the investor's claim; see Christian Tietje and Clemens Wackernagel, "Enforcement of Intra-EU ICSID Awards", *The Journal of World Investment & Trade* 16 (2015), 205, 239.

judicial system, as emphasized by the ECJ in *Achmea*. However, the Union itself has contributed to the negotiation processes of the Energy Charter Treaty.[69] Therefore, the Member States cannot be accused of infringing their duty to Union loyalty when they submit investor-state disputes to arbitral tribunals according to the Energy Charter.[70] Among the criteria developed in *Achmea*, the central criterion relevant for the Energy Charter Treaty is thus the extent to which the Energy Charter arbitration procedures are subject to control by the Member States' courts.[71] According to *Achmea*, such control must guarantee the full effectiveness of Union law. This can be achieved by national courts following the provisions of EU law – as far as applicable in the individual case concerned – and making a reference to the ECJ for a preliminary ruling if necessary.

However, the Energy Charter Treaty does not contain any specific procedural requirements.[72] The investors can rather choose between the different existing international arbitration rules. These are the arbitration rules of the International Centre for Settlement of Investment Disputes (ICSID), the rules of United Nations Commission on International Trade Law (UNCITRAL) or the rules of the Arbitration Institute of the Stockholm Chamber of Commerce (SCC). The UNCITRAL und SCC arbitration rules give the Contracting Parties the right to choose the applicable law and possibly make a reference to the ECJ for a preliminary ruling. This means that an arbitral award can be subject to the law of a Member State, whose courts must comply with Union law and its definite interpretation by the ECJ. This can ensure the autonomy of the Union law and the final decision competence of the ECJ.[73] In such an interpretation, intra-EU arbitration proceedings based on the Energy Charter Treaty are not generally contrary to EU law. The decisive question appears to be whether the national law that is applicable in the individual case allows a full or only limited control of arbitral awards. On the contrary, the arbitration proceedings that rely on the ICSID arbitration rules seem to be problematic from the perspective of EU law.[74] According to Article 54 of the ICSID-Convention,[75] all Contracting

---

69 Gundel, in: *Berliner Kommentar zum Energierecht*, Vol. 1, Einleitung D EnWG, para. 1 et seq.
70 So also Gundel, *Europäisches Wirtschafts- und Steuerrecht* (2018), 124, 128.
71 Cf. European Court of Justice (Grand Chamber), judgment of 6 March 2018, case C-284/16, ECLI:EU:C:2018:158, *Achmea*, para. 50 et seq.
72 Gundel, *Europäisches Wirtschafts- und Steuerrecht* (2018), 124, 127.
73 Cf. also Gundel, *Europäisches Wirtschafts- und Steuerrecht* (2018), 124, 129.
74 Gundel, *Europäisches Wirtschafts- und Steuerrecht* (2018), 124, 129.
75 Article 54 para. 1 ICSID Convention, Regulation and Rules: "Each Contracting State shall recognize an award rendered pursuant to this Convention as binding

States have to enforce the arbitral awards as final judgments, precluding them from further control by the national courts. This, however, conflicts with the ECJ's *Achmea* requirements. Others argue for a restrictive interpretation of Article 26 para. 6 of the Energy Charter Treaty.[76] According to this provision, an arbitral tribunal established to settle disputes between a Contracting Party and an Investor of another Contracting Party "shall decide the issues in dispute in accordance with this Treaty and applicable rules and principles of international law". In order to avoid a conflict between Union law and the Energy Charter Treaty, Article 26 para. 6 of the Energy Charter Treaty should be interpreted restrictively and generally preclude the application of Union law in arbitration procedures as "applicable rules" or as part of "international law". Such an interpretation is considered to be compatible with the ECJ's opinion on CETA.[77] In CETA the ECJ approved the compatibility of arbitration mechanisms established by international agreements which have also been ratified by the EU as compatible with Union law, provided that the arbitral tribunals are not allowed to interpret Union law in a binding way.[78]

The impacts of ECJ's new jurisprudence on investment arbitration relying on the Energy Charter Treaty have not been clarified yet. The aforementioned Draft Agreement for the Termination of BITs between the EU Member States "addresses [only] intra-EU bilateral investment treaties". On the contrary, the agreement "does not cover intra-EU [investor-state arbitration] proceedings on the basis of Article 26 of the Energy Charter Treaty. The Union and its Member States will [rather] deal with this matter at a later stage". However, the European Commission expresses the view that Article 26 of the Energy Charter Treaty does not apply to intra-EU disputes.[79] It is further not clear which legal

---

    and enforce the pecuniary obligations imposed by that award within its territories as if it were a final judgment of a court in that State. [...]".
76 Germelmann, in: *Grundfragen des (Energie-)Regulierungsrechts und des Wirtschaftsrechts*, pp. 110–112.
77 Germelmann, in: *Grundfragen des (Energie-)Regulierungsrechts und des Wirtschaftsrechts*, p. 112.
78 Cf. Ruffert, *Juristische Schulung* (2019), 598, 599.
79 See e.g. Proposed Brief of the European Commission on behalf of the European Union as amicus curiae in support of the Kingdom of Spain before of the United States District Court for the District of Columbia of 3 May 2019, Civil Action No. 1:18-cv-2254, *Masdar Solar & Wind Cooperatief U.A. v. Kingdom of Spain* (Enforcement of an investment arbitration award against Spain based on the Energy Charter Treaty), pp. 11 et seq., https://ec.europa.eu/competition/court/observations.html (accessed 15 April 2020); Rusche, *Europäisches Zeitschrift für Wirtschaftsrecht* (2020), 169, 171.

implications this agreement will have on arbitration procedures between Contracting States according to Article 27 of the Energy Charter Treaty. At the present time, negotiations are taking place at EU level regarding the modernisation of the Energy Charter Treaty.[80] The modernised Energy Charter Treaty should inter alia afford legal certainty for investors and investments of Parties in each other's market.[81] One can hope that the modernised Energy Charter Treaty will be able to resolve the conflict of interest between EU law and international investment arbitration. This requires clear answers, especially concerning the jurisdiction of EU courts, the interpretation and application of EU law by non-EU adjudicatory bodies, and the future of investor-state arbitration within the EU.[82]

Conclusion in Theses

1. The ECJ's *Achmea* ruling leads to the inapplicability of intra-EU BITs.
2. The CETA opinion of the ECJ clarifies that Union law does not generally preclude multilateral arbitration tribunals.
3. Intra-EU arbitration procedures on the basis of the Energy Charter Treaty are highly controversial. In the European Commission's view, intra-EU dispute settlement based on the Energy Charter Treaty is generally contrary to EU law. More convincing is, however, a differentiated approach. Intra-EU investment arbitration can be seen as compatible with EU law, provided that the parties choose a procedural code which is in line with the requirements developed in the *Achmea* judgment.

---

80 European Commission, Recommendation (EU) for a Council Decision authorising the entering into negotiations on the modernisation of the Energy Charter Treaty, COM(2019) 231 final of 14 May 2019; see also Rusche, *Europäische Zeitschrift für Wirtschaftsrecht* (2020), 169, 171.

81 Annex to Council Decision of 2 July 2019, Negotiating Directives for the Modernisation of the Energy Charter Treaty, 10745/19 ADD 1, p. 3; see also Council of the EU, Press release 539/19 of 15 July 2019, https://www.consilium.europa.eu/de/press/press-releases/2019/07/15/council-adopts-negotiation-directives-for-modernisation-of-energy-charter-treaty/ (accessed 15 April 2020).

82 Cf. Proposed Brief of the European Commission on behalf of the European Union as amicus curiae in support of the Kingdom of Spain before of the United States District Court for the District of Columbia of 3 May 2019, Civil Action No. 1:18-cv-2254, *Masdar Solar & Wind Cooperatief U.A. v. Kingdom of Spain*, p. 24.

Prof. Dr. Antonis Metaxas*

2 Safeguarding the autonomy of EU Law: The impact of ISDS mechanisms after Opinion 1/17

Abstract Opinion 1/17 generated substantial scientific debate about the impact of Investor State Dispute Settlement mechanisms on the dynamic notion of the 'autonomy' of the EU legal order. While analysing Opinion 1/17, it is important to evaluate the arguments that convinced the Court to reach the conclusion that the creation of an Investment Court System provided in CETA to handle investment disputes is compatible with EU law. Focus on the merits of these arguments is amplified by the constant efforts of the CJEU to safeguard its strategic position as the sole guardian of the EU Treaties and of the EU legal order as a coherent normative system. The present analysis primarily explores the critical points in the Court's arguments that are related to the notion of autonomy. The primary argument put forth is that the rationale behind Opinion 1/17 leaves an existent, although narrow, risk for the adequate preservation of the autonomy of EU legal order that needs to be addressed. The necessity of a cautious openness of the EU law system to international dispute settlement mechanisms as well as important political considerations underline the need for the CJEU to find, in the future, ways for an *inclusion* of the arbitral dispute settlement structures. Inevitably so, the present analysis highlights the fundamental goal to preserve the autonomy of EU legal order while exploring the pathway to reconcile two necessities: the need for an autonomous 'self-dependence' of the Union's legal system and the need of conciliation in the field of international investment arbitration. The strategic importance of safeguarding the autonomy of EU Law, a *conditio sine qua non* for the EU integration process in general, should better rely on practical, technical ways for its observance than to policy influenced fluctuations of its normative substance.

Keywords: Opinion 1/17, autonomy, preliminary reference procedure, EU law as matter of fact, interpreting EU law

Introduction

The relationship between existing Investor State Dispute Settlement (hereafter 'ISDS') mechanisms and EU law has been an issue of substantial controversy in

---

* Dr. iur. (FU Berlin); Asst. Professor of EU Law, National & Kapodistrian University of Athens / Vis. Professor, TU Berlin.

the past years. In relation to the fundamental principle of autonomy of the EU legal order, the 2018 landmark *Achmea* ruling has been the most debated judgment in this context until Opinion 1/17.[1] The Court of Justice of the European Union (CJEU) handed down its Opinion 1/17 on the compatibility of the Comprehensive Economic and Trade Agreement (CETA) between Canada and the EU with EU law, on the 30th of April 2019.[2] CETA is one of the most recent free trade agreements adopted by the EU, including EUSFTA with Singapore and EUVFTA with Vietnam, that include provisions on investment protection. Opinion 1/17 of the CJEU on CETA keeps the debate on the principle of topical autonomy, claiming that the establishment of the CETA Tribunal and Appellate Tribunal for disputes between investors and the contracting parties is consistent with EU law. In particular, the Court found that the relevant provisions in CETA do not violate the principles of autonomy, equal treatment and effectiveness. The request for an opinion by the Court originated from a fierce dispute within Belgian internal politics, with Wallonia demanding from the Government in Brussels to expressly consult the CJEU on the legal merits of that agreement. Respecting that decision of its regional parliament, Belgium asked the CJEU, inter alia, whether such an agreement was compatible with the principle of autonomy of the EU legal order.

Opinion 1/17 touches upon a number of important and controversial issues. The following analysis does not intend to provide an overview of all the key issues raised by the Court but rather focus predominantly on the approach adopted by the CJEU for safeguarding the autonomy of EU legal order. The primary objective is to identify the elements in the Court's argumentation that could potentially pose a risk for the conceptual and factual outline of the autonomy topos. The basic argument provided is that the reasoning behind Opinion 1/17 raises a narrow but nevertheless existent risk-potential concerning the adequate preservation of EU law self-dependence; a risk that needs to be addressed. At a second level, this analysis argues in favour of the need for the CJEU to find ways for a comprehensive **conceptual inclusion** of the arbitral dispute settlement structures. If something is now clear after Opinion 1/17, it is that autonomy of EU law should better rely on practical, technical ways for the observance of its scope and normative function.

---

1 CJEU (Grand Chamber), Judgment of 6 March 2018, case C-284/16, ECLI:EU:C:2018:158, *Slovak Republic v Achmea BV*.

2 Comprehensive Economic and Trade Agreement (CETA) agreement between Canada, on the one part, and the European Union and its Member States, on the other part, OJ L 11, 14 January 2017, 23.

A practical tool in this direction could be, as it is here argued, a 'smart' use of the preliminary reference procedure provided for in Article 267 TFEU. The substantiation of these arguments can be materialized through the examination of the elements that constitute the *concept* of autonomy of the EU legal order, as well as how these elements have been identified, featured and interpreted before and throughout Opinion 1/17.

1. The concept of autonomy

An independent legal system, should it aim to remain '*independent*' must safeguard its autonomy. Autonomy could be defined as the lack of normative **control** from external sources as regards (at least) the central structural decisions it entails and the values it reflects. In this sense, safeguarding its autonomy is not an 'egoistic' perception and tendency but rather a precondition of the very existence of a given coherent legal order. Considering that the EU, seen as a project with a mainly political *telos*, is founded on the legitimation of a distinctive and autonomous legal order[3], the concept of autonomy constitutes a structural existential principle which is inextricably linked to the present development as well as the further evolution of the European integration process.[4] Beyond 'merely' a system with primacy over the laws of the Member States, the principle of autonomy of EU law essentially ensures that the common set of rights and obligations deriving from the Treaties to form the EU legal order will be sheltered from external factors that would undermine its coherence. The normative substance of the autonomous EU legal order takes its more concrete form precisely at the extreme crucial constellation, when there is a genuine *collision* with national and/or international law.[5] In addition to the above, it needs to be underlined that the existential cornerstone of the EU legal order, the

---

3   Antonis Metaxas, "Reflections on the distinctive character of the EU legal order", *Efimerida Dioikitikou Dikaiou* (2016), Vol. 3, 346–355.
4   Niamch Nic Shuibhne, "What is the autonomy of EU law, and why does that matter?", *Nordic Journal of International Law* (2019), Vol. 88, 9–40; on the principle of autonomy see also Bruno de Witte, "European Union Law: How autonomous is its legal order?", *Zeitschrift für Öffentliches Recht* (2010), 141; Jed Odematt, "The Principle of Autonomy: an Adolescent Disease of EU External Relations Law?", in: *Structural Principles in EU External Relations Law*, ed. Marise Cremona (Oxford and Portland: Hart Publishing, 2018), p. 291.
5   Niamch Nic Shuibhne, op. cit., 4; see also Antonis Metaxas, "State of Exception" as the new *legitimitas*: Preliminary observations on the necessity of an interdisciplinary approach in EU Law", *Efimerida Dioikitikou Dikaiou* (2018), 642.

supremacy principle, is also predominantly based on the central assumption of structural autonomy of the EU legal order. Since the *Costa* case, the CJEU's jurisprudence has highlighted the principle of supremacy as the *key* methodological tool of conflict resolution. The legal consequence of the principle of supremacy is the inapplicability of national rules that are in conflict with EU law provisions (*Anwendungsvorrang*).[6]

The notion of autonomy has been particularly constructed in the Court's case law. The pivotal role of the CJEU is illustrated in many respects, as the Court is acting as the 'guardian' of the EU normative framework, under the EU's *sui generis* status. Historically, the CJEU did not hesitate to act as an activist court that *constructed* and *ab initio* formed, to a large extent, the dogmatic pillars of the EU legal order even in cases where those pillars did not have a clear foundation in the Treaties.[7] The Court founded the approach of the EU as a *Rechtsgemeinschaft*, a *'community of law'*, whose dogmatic constitution is based on a sequence of interrelated theoretical doctrines and procedural mechanisms: supremacy of EU law, direct effect and the principle of state liability for breaches of EU law, are the most symbolic fundamental principles based on which the EU legal order has not just been *shaped* but indeed *constructed*.[8] In the landmark judgment *Van Gend en Loos,* the CJEU claimed that what is at stake regarding the principle of autonomy is the control or monopoly of jurisdiction of the Court aiming to protect the essential characteristics of the EU and its legal order. This approach of the Court inevitably leads to the necessity to identify and analyse the ways and forms in which CJEU's monopoly of jurisdiction is manifested and legitimized, thus identify the various aspects of autonomy itself.

---

6 Antonis Metaxas, op. cit. 3; see also CJEU, judgment of 15 July 1964, case 6/64, ECLI:EU:C:1964:66, *Flaminio Costa v. E.N.E.L.*

7 CJEU, Judgment of 15 July 1964, case 6/64, ECLI:EU:C:1964:66, *Flaminio Costa v. E.N.E.L.*

8 CJEU, Judgment of 5 February 1963, case 26/62, ECLI:EU:C:1963:1, *Van Gend en Loos v. Administratie der Belastingen*; *Flaminio Costa v. E.N.E.L.*, op. cit.; CJEU, judgment of 19 November 1991, joined cases C-6/90 and C-9/90, ECLI:EU:C:1991:428, *Andrea Francovich and Danila Bonifaci and others v Italian Republic*; see also Michael Dougan, "The Vicissitudes of Life at the Coalface: Remedies and Procedures for Enforcing Union Law before the National Courts", in: *The Evolution of EU Law*, eds. Gráinne De Búrca and Paul Craig (Oxford: Oxford University Press, 2011).

2. The fundamental aspects of the principle of autonomy in the light of Opinion 1/17

There are three dominant criteria structuring the principle of autonomy and highlighting its legitimacy and necessity, namely the allocation of competences between the EU and the Member States, the importance of the preliminary reference mechanism, and the normative control *of* and *on* EU law.[9]

2.1 Allocation of competences

The first aspect of autonomy is about the division of competences between the Union and Member States. Essentially, this refers to the definition of the scope of the sovereignty of the Member States in the field of law making. On this issue, Opinion 1/17 seems to offer solid ground on the argument that provisions in CETA essentially offer CJEU undisputed monopoly of jurisdiction for the determination of the division of competencies. Already in Opinion 1/91 the Court identified autonomy in terms of protection against adverse effects on the allocation of responsibilities defined in the Treaties, while special emphasis was placed on its own competence both to articulate and to assure respect for that definition.[10] The power to effectively control external interaction is therefore highly concentrated on the Court. This dimension is clearly reflected in the jurisprudence of the CJEU. In Opinion 1/91, the Court claimed that if the EEA Court could be called upon to interpret the expression 'Contracting Party', then the autonomy would be breached as it "is likely adversely to affect the allocation of responsibilities defined in the Treaties and, hence, the autonomy of

---

[9] Francisco de Abreu Duarte, "Autonomy and Opinion 1/17 – a matter of coherence", *European Law Blog*, 31 May 2019, https://europeanlawblog.eu/2019/05/31/autonomy-and-opinion-1-17-a-matter-of-coherence/ (accessed 4 February 2020); see also Bruno de Witte, "Direct Effect, Primacy, and the Nature of the Legal Order", in: *The Evolution of EU Law*, eds. Paul Craig and Grainne De Burca (Oxford: Oxford University Press, 2011); Szilárd Gaspar-Szilágyi, "A Standing Investment Court under TTIP from the Perspective of the CJEU", in: *Journal of World Investment and Trade* (2019), 701–742; Jed Odermatt, "When a fence becomes a cage: the principle of autonomy in EU external relations law", *EUI Working Papers*, https://cadmus.eui.eu/bitstream/handle/1814/41046/MWP_2016_07.pdf?sequence=1 (accessed 4 February 2020); Piet Eeckhout, "Human rights and the autonomy of EU law: pluralism or integration?", *Current Legal Problems* (2013), Vol. 66, Issue 1, 169–202; Inge Govaere and Sacha Garben, "The Interface Between EU and International Law: Contemporary Reflections" (Oxford: Hart Publishing, 2019).

[10] CJEU, Opinion 1/91 of 14 December 1991, ECLI:EU:C:1991:490, para. 2.

the Community legal order, respect for which must be assured by the Court of Justice pursuant to Article 164 of the EEC Treaty".[11]

Furthermore, in Opinion 2/13 the Court stated that "However, the fact remains that, in carrying out that review, the ECHR would be required to assess the rules of EU law governing the division of powers between the EU and its Member States as well as the criteria for the attribution of their acts or omissions, in order to adopt a final decision in that regard which would be binding both on the Member States and on the EU. [. . .] Such a review would be liable to interfere with the division of powers between the EU and its Member States".[12] Last, it should be also noted that the Court has constantly defended its monopoly to declare an unlawful act of EU law to be void (Case 314/85, *Foto-Frost*), this being also an expression of this profound fundamental conception: the autonomy of the EU legal order.[13] In Article 8.21 of CETA on the determination of the respondent for disputes, it is stated that "the European Union shall, after having made a determination, inform the investor as to whether the European Union or a Member State of the European Union shall be the respondent".[14] In Opinion 1/17, the Court makes a clear mention on the significance of this element for the autonomy of EU law. The Court argues that, by explicitly providing the power to determine whether a possible dispute should be brought against a Member State or against the Union and is granted on the Union and not on the CETA Tribunal, the exclusive jurisdiction of the Court to give rulings on the division of powers between the Union and its Member States is preserved.[15]

---

11 Ibid., para. 34–35; Francisco de Abreu Duarte, op. cit. 9, refers also to the *Mox Plant* case, where the Court stated: "The act of submitting a dispute of this nature to a judicial forum such as the Arbitral Tribunal involves the risk that a judicial forum other than the Court will rule on the scope of obligations imposed on the Member states pursuant to Community law".
12 CJEU, Opinion 2/13 of 18 December 2014, ECLI:EU:C:2014:2454, para. 224–225.
13 See Jürgen Bast, "Autonomy in Decline? A commentary on Rimšēvičs and ECB v Latvia", *Verfassungsblog*, 13 May 2019, https://verfassungsblog.de/autonomy-in-decline-a-commentary-on-rimsevics-and-ecb-v-latvia/ (accessed 3 February 2020), with reference to the very important recent judgment of the Court in the Rimšēvičs case where the CJEU declared for the first time a national legislative act of a Member State void.
14 Article 8.21, Comprehensive Economic and Trade Agreement (CETA) of 14 January 2017, between Canada, the European Union and its Member States.
15 CJEU, Opinion 1/17 of 30 April 2019, ECLI:EU:C:2019:341, *CETA*, para. 13.

2.2 The significance of the preliminary reference procedure

The second crucial aspect of autonomy as well as an indispensable tool for safeguarding its essence is the respect for the mechanism of preliminary reference that safeguards the fundamental link of the Court with national courts. This link is a *conditio sine qua non* for the strategic structural task assigned to the CJEU under Article 267 TFEU, the object of which is to secure uniform interpretation of the Treaty by national courts and tribunals.[16] To this end, the responsibilities but also the privileges of national courts and tribunals to ensure their functioning as EU courts within that system must be protected.[17] Legal redress for the individual is thus safeguarded and further homogenous evolution of EU law is guaranteed through the preliminary ruling procedure.[18] Such considerations are present in the *Achmea* case, thus hinting towards a connection between the principle of autonomy of EU law and the rule of law.[19] Opinion 1/17 seems – at a first glance – to be departing from the requisitions of this element, since CETA does not provide for any function that could simulate a system of preliminary ruling in the Investment Court System it introduces. The CJEU however, assesses that this does not pose a threat to the application of EU law due to the way it –EU law – is described in CETA. In several occasions in Opinion 1/17, the Court of Justice refers to Article 8.31.2 of the CETA stipulating that "the Tribunal will have to confine itself to an examination of EU law 'as a matter of fact' and will not be able to engage in interpretation of points of law".

Overall, in contrast to traditional regional or international courts, the CJEU is not only assigned with the application, interpretation and validation of

---

16 CJEU, Judgment of 5 February 1963, case 26/62, ECLI:EU:C:1963:1, *Van Gend en Loos v. Administratie der Belastingen*, para.7; see also CJEU, Opinion 2/13 of 18 December 2014, ECLI:EU:C:2014:2454, para. 176.
17 Niamch Nic Shuibhne, op. cit. 5.
18 Joseph H.H. Weiler, "Revisiting Van Gend en Loos: Subjectifying and Objectifying the Individual", in: *50th Anniversary of the Judgment in Van Gend en Loos 1963-2013*, eds. Antonio Tizzano et al. (Luxembourg: Office des publications de l'Union européenne, 2013), p. 21, curia.europa.eu/jcms/jcms/P_95693 (accessed 3 February 2020); see also Carlos Closa, Dimitry Kochenov and Joseph H. H. Weiler, "Reinforcing rule of law oversight in the European Union", *Robert Schuman Centre for Advanced Studies Research Paper* (2014), Vol. 25.
19 See also Steffen Hindelang, "Conceptualisation and Application of the Principle of Autonomy of EU Law – The CJEU's Judgement in Achmea Put in Perspective", *European Law Review* (2019), Vol. 3, 383-400.

secondary EU law provisions and with the interpretation of primary EU law, but it has also established itself as a constitutional-type court. In its latter capacity, well before Opinion 1/17, the Court has developed principles and mechanisms (primacy, direct effect) to define the relationship between the EU and Member State legal orders. The Court has been engaged in a constant dialogue with the Member State courts through the preliminary reference mechanism under Article 267 TFEU. More importantly, the Court has created an intricate case-law on the relationship between the EU legal order and international law.[20] The preliminary ruling procedure is therefore inherently linked to the autonomy of EU law, being described as "essential" and "indispensable to the preservation of the very nature of European Union law".[21] This statement can be originally found in Opinion 1/09 on the establishment of a European and Community Patents Court, and provides the principal argument for the CJEU so as not to allow the possibility for such a court to ignore domestic courts and acquire exclusive jurisdiction over that part of EU law. CJEU underlined that "the tasks attributed to the national courts and to the Court of Justice respectively are indispensable to the preservation of the very nature of the law established by the Treaties".[22] In 2018, in the landmark *Achmea case*, the significant role of the preliminary reference procedure was highlighted extensively. According to the CJEU, arbitral courts could not be seen as courts in the sense of Article 267 TFEU, as they stood outside the EU's legal system and could not interpret EU law.[23]

In *Achmea,* the Court argued that "the judicial system as thus conceived, has as its keystone the preliminary ruling procedure provided for in Article 267 TFEU, which, by setting up a dialogue between one court and another, specifically between the Court of Justice and the courts and tribunals of the Member states, has the object of securing uniform interpretation of EU law, thereby serving to ensure its consistency, its full effect and its autonomy as well as, ultimately, the particular nature of the law established by the Treaties".[24]

---

20 Szilárd Gaspar-Szilágyi, op. cit. 9; see also Bruno De Witte, "Direct Effect, Primacy, and the Nature of the Legal Order", in: *The Evolution of EU Law*, eds. Paul Craig and Grainne De Burca (Oxford: Oxford University Press, 2011), pp. 346–348.
21 CJEU, Opinion 1/09 of 8 March 2011, ECLI:EU:C:2011:123, para. 3, 89.
22 Ibid, para. 85.
23 Jens Hillebrand Pohl, "Intra-EU Investment Arbitration after the Achmea Case: Legal Autonomy Bounded by Mutual Trust?", *European Constitutional Law Review* (2018), Vol.14, Issue 4, 767–791.
24 CJEU (Grand Chamber), Judgment of 6 March 2018, case C-284/16, ECLI:EU:C:2018:158, *Slovak Republic v Achmea BV*, para. 37.

In addition, as aptly mentioned by Biltgen "the scope of *Achmea* is essentially limited to arbitration clauses in BITs between Member States and does not destroy bridges between the Courts of the EU and those of Member States"[25]. Regarding the future of investment treaty arbitration, the *Achmea* ruling may urge Member States to terminate their intra-EU BITs, even though most such BITs provide for 'sunset clauses' – an extended period of applicability following termination.[26] However, it has often been argued that autonomy means different things in different contexts, therefore its practical implementation remains unclear.[27] For example, it has not been clarified yet what autonomy means in relation to the Energy Charter Treaty and arbitral tribunals have consistently refused to accept the relevance of *Achmea* in that context.[28] However, intra-EU investment arbitration based on the ECT seems to have the capacity to generate distressful conditions with the possibility of conflicting obligations originating on the one hand from EU law and on the other hand from an arbitral award based on the ECT when EU Member States act as respondents. This situation is practically similar to *Achmea*, regardless the obvious differentiation of the EU being also a party to the ECT together with each Member State.[29] It needs to be noted that the debate on the role of EU public policy in arbitration, when confronted with the recent discussion on the potential inclusion of ISDS in EU

---

25 See François Biltgen, "The concept of autonomy of EU law: from Opinion 2/13 (accession to the ECHR) to Achmea and Opinion 1/17 (CETA)", in: *Building bridges: central banking law in an interconnected world. ECB Legal Conference 2019* (2019), Part 3, 80–89, https://www.ecb.europa.eu/pub/pdf/other/ecb.ecblegalconferencepr oceedings201912~9325c45957.en.pdf?258d648ffcf1be39f9d927e5c13f393f (accessed 3 February 2020).
26 Ibid.; see also Steffen Hindelang, op. cit. 19.
27 See Cristina Contartese, "Achmea and Opinion 1/17: Why do intra and extra-EU bilateral investment treaties impact differently on the EU legal order?", in: *Legal Working Paper Series. The new challenges raised by investment arbitration for the EU legal order* (2019), No. 19, https://www.ecb.europa.eu/pub/pdf/scplps/ecb. lwp19~e4d0a59cea.en.pdf (accessed 3 February 2020).
28 Panos Koutrakos, "More on Autonomy – Opinion 1/17 (CETA)", *European Law Review* (2019), Vol. 3, 293–294, https://dialnet.unirioja.es/servlet/articulo?codigo=6976614 (accessed 3 February 2020).
29 Andrea Pinna, "The Incompatibility of Intra-EU BITs with European Union law, annotation following ECJ, 6 March 2018, Case 284/16, Slovak Republic v Achmea BV", *Paris Journal of International Arbitration, Cahiers de l' arbitrage* (2018), Vol. 1, 73–95; Juliane Kokott and Christoph Sobotta, "Investment Arbitration and EU Law", *Cambridge Yearbook of European Legal Studies* (2016), Vol. 18, 7–8.

investment and trade agreements, does entail proposals to soften EU procedural law in the field of preliminary reference procedure under Article 267 TFEU to allow arbitral panels to seek preliminary rulings before the CJEU.[30] In particular, as it has been concluded in several decisions, ISDS arbitration tribunals acting under a BIT of a Member State would be entitled to request the Court of Justice for preliminary rulings where the claimant investor had the alternative option to bring its case to a national Court.[31] In the *Achmea* case however, the Court raised the question with regard to the necessary mechanisms that would ensure the uniform and consistent interpretation of Union law as EU law formed part of the applicable law. In assessing whether an *ex ante* mechanism (the investment tribunal seeking a preliminary ruling from the CJEU) or an *ex post* control (the investment tribunal's award being reviewed by a domestic court in the EU) could constitute such mechanisms, the Court held that neither approach was legally feasible or, at the end, satisfactory.[32]

Returning to Opinion 1/17, the provision in CETA regarding the Tribunal's position to confine itself to an examination of EU law 'as a matter of fact' seems to be the critical point in CJEU's rationale. In this provision the CJEU finds sufficient support and legitimacy in its objective that the Investment Court System is not given the competence to interpret EU law, thus such a competence remains an exclusive privilege of the CJEU.[33] The same confidence is shared by the Opinion of Advocate General Yves Bot delivered on 29 January 2019. On the CETA provision that the Tribunal may consider the domestic law of a Party as 'a matter of fact', the Advocate General states that "consideration of the Parties' domestic law must not entail the CETA Tribunal amending that law. It must take account of that law as it stands".[34] AG Bot makes a comparison to the *Achmea* judgment stating that "unlike in the case of bilateral investment treaties between Member States such as that at issue in the case which gave rise to the judgment in *Achmea*, EU law does not form part of the international law applicable between the Parties".[35] Opinion 1/17 is therefore resting on the

---

30 Barbara Alicja Warwas, "The State of Research on Arbitration and EU Law: Quo Vadis European Arbitration?", *EUI Department of Law, Research Paper* (2016), Vol. 23.
31 Jürgen Basedow, "EU law in international arbitration: Referrals to the European Court of Justice", *Journal of International Arbitration* (2015), Vol. 32, Issue 4, 367–386.
32 See with further references Cristina Contartese, op. cit. 27.
33 Ibid.
34 Opinion 1/17 of AG Bot delivered on 29 January 2019, ECLI:EU:C:2019:72, *CETA*.
35 Ibid., para. 110.

element that the ICS cannot interpret EU law and, consequently, on the notion that by considering the domestic law 'as a matter of fact', the ICS must follow the prevailing interpretation given by the courts or authorities accepted by the institutions or the courts of the European Union. Further protection against the possibility of misinterpretation of EU law seems to be entrusted in the establishment pursuant to Article 8.28.1 of the CETA of an Appellate Tribunal set to review awards rendered by the ICS. This is based again on the notion that the Appellate Tribunal will be taking EU law into consideration as 'a matter of fact'. In this context, the Appellate Tribunal can revise or overturn an award of the CETA Tribunal on the basis of "manifest errors in the appreciation of the facts, including the appreciation of relevant domestic law".[36] It can be assumed that such a provision is meant as an additional reassurance against the possibility of an error by the Tribunal in its appreciation of the relevant domestic law, that would then be corrected through the process of the review of its awards by the Appellate Tribunal.

The issue that arises by the aforementioned analysis relates directly to the critical element of EU law being considered as 'a matter of fact' -an element through which the CJEU builds its assurances on the autonomy of EU legal order- being used as a tool to secure the proper function of both the CETA Tribunal and Appellate Tribunal. Keeping in mind that the concept is new in CETA, a structure in which the Appellate Tribunal will review an ICS award on its merit to take account of EU law 'as it stands', entails a specific risk: The Appellate Tribunal in order to deliver its judgment on the ICS award, risks eventually interpreting EU law thus deconstructing the fundamental argumentation behind Opinion 1/17. It needs to be noted, however, that there are cases where investment tribunals need to come to an interpretation of domestic law, when for example it needs to be clarified whether a contract was lawfully dismissed under domestic law. A typical example was *Malicorp v. Egypt* where such a clarification was necessary in order to confirm whether any rights susceptible of expropriation persisted.[37] In order, for example, to decide whether an expropriation was done 'under due process of law', as Article 8.12(1)(b) requires, a tribunal may need to examine if the State complied with domestic legal procedures when expropriating the investor. Such an assessment of a local court judgment's consistency with domestic law could be considered as partial interpretation of the local court's engagement in abuse of domestic law,

---

36 Ibid., para. 27.
37 ICSID, award of 7 February 2011, case no. ARB/08/18, *Malicorp Limited v. The Arab Republic of Egypt*.

leading to a denial of justice and a breach of CETA Article 8.10(2)(a). It should therefore be considered that however slim, the possibility of infringement of Article 8.31.2 of the CETA does exist.

2.3 Control over the uniform application of EU law

A third aspect linked to autonomy relates to the level of control the CJEU holds over the proper application of EU law in case a dispute settlement body misinterprets and/or violates EU law. Member States can be on the receiving end of violation measures imposed by the Commission should the CJEU find that a domestic court failed to uphold EU law. Following the insofar analysis, Opinion 1/17 needs to be evaluated on this critical aspect too. In his Opinion AG Bot comments that "infringement of Article 8.31.2 of the CETA would constitute an error in the application of applicable law"[38]. AG Bot identifies such an infringement in the event the Tribunal would end up formulating its own interpretation of EU law, without considering the interpretation of that law as accepted by the institutions or the courts of the European Union. Furthermore, AG Bot emphasises that a review by the Appellate Tribunal should be conducted only in the event that there is nothing in the EU legal order to clarify the meaning to be given to a provision of EU law.

In Opinion 1/17 we find that the CJEU responds to the question by emphasizing that such an examination by the Appellate Tribunal could not be considered as the equivalent to an interpretation of domestic law. Again, the Court's argumentation is based on the notion that in such a case the domestic law would be taken 'as a matter of fact' thus the risk of ending up with an interpretation by the CETA Tribunal does not exist. It needs to be noted that the aforementioned argumentation essentially denies the possibility of infringement of CETA Article 8.31.2 rather than stipulating – or even indicating – on the suggested action taken in case of such an infringement. So far, in cases where national courts in Member States fail to uphold EU law, either by not complying with the interpretation offered by the CJEU following an answer to a preliminary ruling, or by not requesting a mandatory preliminary ruling in the first place, the CJEU retains its competence to submit corresponding sanctions.[39] Not only that: According to the

---

38 Opinion of AG Bot, op cit., para. 154.
39 Joana Covelo de Abreu, "Infringement Procedure and the Court of Justice as an EU Law's Assurer: Member States' Infringements Concerning Failure to Transpose Directives and the Principle of an Effective Judicial Protection", in: *Towards a Universal Justice? Putting International Courts and Jurisdictions into Perspective*, ed. Dário Moura Vicente (Leiden: Brill Nijhoff Publishing, 2016), pp. 468–475, https://

famous *Köbler* judgment, Member States are obliged to compensate the damage caused to individuals in cases where an infringement of EU law stems from a decision of a Member State court adjudicating in the last instance[40]. In the landmark judgment *Commission v France* rendered in 2018, the CJEU condemned for the first time a Member State for a breach of Article 267(3) TFEU in the context of an infringement action, after the French administrative Supreme Court (Conseil d'Etat) failed to make a necessary preliminary reference. This decision is undoubtedly a determining step towards a more complete system of safeguards put by the Court so that it could be able to remain in full control of the system of the EU legal order as a whole. All these instruments cannot be activated in the case of tribunals that are not competent to apply EU law like the Investment Court System (ICS) provided for by CETA as described in the following analysis (see below under IV).

Corresponding examples related to this issue can be found in Opinion 1/09 where it is stated that "…it is clear that if a decision of the Patent Court were to be in breach of European Union law, that decision could not be the subject of infringement proceedings nor could it give rise to any financial liability on the part of one or more Member State…".[41] This is due to the fact that the European and Community Patents Court (PC) is not attached to a Member State that could be held responsible for the infringement of EU law. Similar concerns rose, in CJEU's opinion 2/13 regarding the procedure established by Protocol No 16 of the Convention for the Protection of Human Rights and Fundamental Freedoms. The Protocol provided for the highest courts and tribunals of the Contracting Parties to be able to request the ECHR to give advisory opinions on questions of principle related to the interpretation or application of the rights and freedoms defined in the ECHR or the Protocols thereto. CJEU expressed its concerns that the procedure established by Protocol No 16 may apply "even though EU law requires those same courts or tribunals to submit a request to that end to the Court of Justice for a preliminary ruling under Article 267 TFEU".[42] In Opinion 1/17 however, CJEU's perspective on the issue seems to alter from its previous concerned standpoint. The Court accepts that CETA

---

brill.com/view/book/edcoll/9789004298712/B9789004298712-s028.xml (accessed 3 February 2020).
40   CJEU, Judgment of 30 September 2003, case C-224/01, ECLI:EU:C:2003:513, *Gerhard Köbler v. Republik Österreich*; see Antonis Metaxas, "Member State Liability for judicial breaches of EU Law", *Efimerida Dioikitikou Dikaiou* (2013), Vol. 5, 727–736.
41   CJEU, Opinion 1/09 of 8 March 2011, ECLI:EU:C:2011:123, para. 88.
42   CJEU, Opinion 2/13 of 18 December 2014, ECLI:EU:C:2014:2454, para. 196.

Article 8.28.2(b) allows the Appellate Tribunal to identify possible errors in the appreciation of relevant domestic law. However, the CJEU rests confident on the concept that preceding provisions in CETA offer assurance that the intention of the Parties to the agreement was not to attribute jurisdiction to the Appellate Tribunal to interpret domestic law.[43] It is therefore evident that while with Opinion 1/17 the Court does not change its standpoint on the importance of retaining its competence to submit corresponding sanctions in case of infringement, it accepts that in CETA such an infringement is simply not possible, thus no concerns need to be raised. However indicative of the Court's intention to offer a 'softer' approach towards its safeguarding tone on the issue of control of EU law, it needs to be highlighted that such an approach is not without risk. The simple notion that the intention of the Parties is not to confer jurisdiction to the Appellate Tribunal to interpret domestic law, does not actually avert the possibility of such a development. In such a case, Opinion 1/17 offers no clarification neither on what would follow such an infringement nor on what would that -in essence- mean for its control of EU law.

3. Evaluating the principle of autonomy after Opinion 1/17

Out of the three aspects of autonomy already described, the significance of the mechanism of preliminary reference as a fundamental element structuring the relationship between the CJEU and domestic courts has been intensively outlined. Opinion 1/17 could come here as a surprise, since, in essence, it opens the possibility for the Investment Court System (ICS) provided by CETA to override the mechanism of preliminary reference. It needs, however, to be noticed that it was the Commission itself that first came with a proposal on the integration of arbitration with the EU legal regime, a proposal that although not eventually materialized, did renew the scientific debate on the need to combine arbitration within the EU procedural law system. In its 2015 Communication to the European Parliament, the Commission declared its resolve to ensure that "EU bilateral agreements will begin the transformation of the old investor–state dispute settlement into a public Investment Court System composed of a Tribunal of first instance and an Appeal Tribunal operating like traditional courts". The Commission proposed to engage in an effort with other international partners to build consensus for a fully-fledged, permanent International

---

43 CJEU, Opinion 1/17 of 30 April 2019, ECLI:EU:C:2019:341, *CETA*, para. 131, 133.

Investment Court and to support the incorporation of investment rules into the WTO.

According to the Commission, such action could lead to a clear code of conduct to avoid conflicts of interest, while engaging "independent judges with high technical and legal qualifications comparable to those required for the members of permanent international courts, such as the International Court of Justice and the WTO Appellate Body" and offering an opportunity to "simplify and update the current web of bilateral agreements and to set up a clearer, more legitimate and more inclusive system".[44] The Commission's proposal is indicative of the pressuring necessity to establish a connection between investor-state dispute settlement bodies and the EU procedural legal order.[45] Discussion, on the position EU public policy holds towards ISDS in EU investment agreements should be expected to eventually result in a framework that eases EU procedural law in the field of preliminary reference procedure to allow arbitral tribunals to seek preliminary rulings before the CJEU. Opinion 1/17 should be considered as a ruling that will undoubtedly impact negotiations on ISDS going forward well beyond Europe. At the ongoing negotiations at UNCITRAL Working Group III, EU is indeed proposing the establishment of a permanent multilateral investment court with an appeal mechanism and full-time adjudicators as the only reform option that can effectively respond to all the concerns on multilateral reform of ISDS. CETA's investment dispute settlement mechanism could constitute the basis for potential bilateral agreements to which the EU is party.[46] However, it needs to be stressed that, on the issue of the interpretation of EU Law, under the CETA provisions, the "risk" of the Appellate Tribunal eventually de facto ending up interpreting EU law, is still present. On this critical issue, as presented earlier in this analysis, CJEU's argument in Opinion 1/17 is that while Article 8.28.2(b) of the CETA offers the Appellate Tribunal the ability to identify possible errors in the appreciation of relevant domestic

---

44 European Commission, Trade for All: Towards a more responsible trade and investment policy, COM/2015/0497 final of 14 October 2015, https://ec.europa.eu/transparency/regdoc/rep/1/2015/EN/1-2015-497-EN-F1-1.PDF (accessed 3 February 2020).

45 George A. Bermann, "Reconciling European Union Law Demands with the Demands of International Arbitration", *Fordham International Law Journal* (2011), Vol. 34, Issue 5, 1197.

46 See further on Marc Bungenberg and Catharine Titi, "CETA Opinion – Setting Conditions for the Future of ISDS", *Blog of the European Journal of International Law*, 5 June 2019, https://www.ejiltalk.org/ceta-opinion-setting-conditions-for-the-future-of-isds/ (accessed 3 February 2020).

law, the preceding provisions make it clear that it was not the *intention* of the Parties to confer on the Appellate Tribunal jurisdiction to interpret domestic law. Referring simply to the *intention* of the Parties does not seem to offer adequate argumentation against the aforementioned risk. This, in turn, leads to a final issue raised by Opinion 1/17 regarding the provision that "any meaning given to domestic law by the Tribunal shall not be binding upon the courts or authorities of that Party".[47] Although such a provision solves the issue of not changing the nature of EU law or case-law for the CJEU, it does not eliminate the possibility of exposing the Member States to conflicting situations between the ICS and CJEU. In the – slim but not eliminated – possibility that the ICS makes a "mistake" while interpreting domestic law 'as a matter of fact', the respective MS would be exposed to rulings by the ICS for acts which could even have been enacted by imposition of EU law, while on the other hand the CJEU could impose sanctions to a MS that would be leading to legislative changes in order to comply with rulings by the ICS triggered by Canadian investors.

Conclusions

Opinion 1/17 has rightfully generated substantial scientific debate regarding the conceptions of autonomy of EU law and their correlation with Investor State Dispute Settlement mechanisms. Opinion 1/17 provides an interesting line of argumentation based on which the Court was convinced in reaching the conclusion that the creation of an Investment Court System as provided in CETA to handle investment disputes is compatible with EU law. Focus on the merits of these arguments is amplified by the efforts of the CJEU to safeguard its central position as the guardian of the EU Treaties. Taking the normative delineation of the autonomy of EU legal order into account, it seems that the critical point that holds the CJEU rationale together is its resolve that the Tribunal provided by CETA will have to confine itself to an examination of EU law 'as a matter of fact'. This concept seems to be the key in Opinion 1/17 since by taking EU law 'as a matter of fact', the ICS provided in CETA will neither be able to engage in interpretation of points of law nor make awards that might have the effect of preventing the EU institutions from operating in accordance with the EU's constitutional framework. According to the Court, these two elements safeguard that the autonomy of EU legal order is preserved. Regardless of the Court's resolve though, concerns could be raised with regard to the small – but

---

47   CJEU, Opinion 1/17 of 30 April 2019, ECLI:EU:C:2019:341, *CETA*, para. 130.

nevertheless present – possibility of the Appellate Tribunal interpreting EU law, as well as the CJEU's limited ability to control the situation in case of an infringement of CETA's Article 8.31.2. The Court's resolve that the mere *intention* of the Parties as regards the Appellate Tribunal's jurisdiction to interpret domestic law is adequate safeguard against the aforementioned risk, could be interpreted as an indication of the CJEU's inclination towards a more flexible stance with respect to the standards under which autonomy of EU legal order is preserved. One should expect that these concerns will be the topic of substantial scientific debate as well as further Opinions and judgments by the CJEU in the future.

Wiegand Laubenstein*

3 Judicial settlement of disputes in the energy sector in Germany

Abstract: This chapter offers brief information about the history of German regulatory law, about the Higher Regional Court of Düsseldorf (OLG Düsseldorf) and about three important current cases the OLG Düsseldorf had to deal with. At first, the history of German regulatory law is outlined. The Energy Industry Act (EnWG) has been developed to its current version under the affection of the European legislation. Today the grid sector is a strongly regulated economy sector, whose legal framework is repeatedly changing due to the European and German legislation. The OLG Düsseldorf is a nationwide responsible Court for appeals against decisions of the German Federal Network Agency (BNetzA) in the field of grid-bound energy supply. Therefore, it is an important court, which has to deal with the latest and essential topics in its decisions. In a next chapter, the essay discusses the decisions of the OLG Düsseldorf and the Federal Court of Justice (BGH) on the *Lichtblick*-case. Both courts had to deal with the the question, if the decision of the BNetzA defining the allowed equity capital interest rates for network operators can also lead to a formal and substantive complaint of network users. Furthermore, the essay deals with the incentive regulation of the network operators, in particular the so-called Xgen or X-factor as a part of the incentive regulatory formula. The calculation of the Xgen is highly controversial . In July 2019, the OLG Düsseldorf dealt with the decision of the BNetzA regarding the determination of the Xgen for the third regulatory period in the gas sector. The judgments of the OLG Düsseldorf are currently subject of proceedings before the BGH. Finally, the text discusses the decisions of the BNetzA and the European Commission to exempt the Ostsee Pipeline Anbindungsleitung (OPAL) from regulation. These decisions could possibly harm the principals of energy solidarity and energy security, which are guaranteed on a European level.

Keywords: history of German regulatory law, current judicature of the Higher Regional Court of Düsseldorf in energy law, formal and substantive complaint (the *Lichtblick*-case) – X-factor and incentive regulation – principles of energy solidarity and energy security (the OPAL-case)

1 The history of German regulatory law

The history of German regulatory law began in 1996. With the first Electricity Internal Market Directive of 1996, the opening of monopolies to competition in

---

* Former Chairman Judge of the 3rd Cartel Senate of the Higher Regional Court of Düsseldorf.

the energy sector began.[1] The first Gas Directive followed in 1998.[2] The Energy Industry Act (EnWG 1998) had already abolished territorial agreements and closed down supply areas in the electricity market. However, effective competition only arose as a result of the European Union's acceleration directives.[3] The second Directive in 2003 had been the basis for the liberalisation of the markets upstream and downstream of network operation.[4] These acceleration directives had introduced real competition in the electricity and gas networks. The natural monopolies have been subject to the regulation of charges and unbundling.[5] By now, the core objectives of the directives have been achieved, which means the elimination of monopolies, the unbundling of activities, non-discriminatory access to networks and the eligibility of customers. The 2003 directives were replaced by Directives 2009/72/EC and 2009/73/EC of the European Parliament and of the Council of the European Union on the 13th of July 2009 as part of the EU's Third Energy Package.[6] Additionally, Regulation (EC) 1228/2003 was

---

1 Jochen Mohr, "Sicherung der Vertragsfreiheit durch Wettbewerbs- und Regulierungsrecht" (Tübingen: Mohr Siebeck, 2015), p. 551; Franz Jürgen Säcker, in: *Berliner Kommentar zum Energierecht,* ed. Franz Jürgen Säcker (Frankfurt am Main: Fachmedien Recht und Wirtschaft, 2019), Vol. 1, Einl. A. rec. 21; Directive 96/92/EC of the European Parliament and of the Council of 19 December 1996 concerning common rules for the internal market in electricity.
2 Directive 98/30/EC of the European Parliament and of the Council of 22 June 1998 concerning common rules for the internal market in natural gas.
3 Directive 2003/54/EC of the European Parliament and of the Council of 26 June 2003 concerning common rules for the internal market in electricity and repealing Directive 96/92/EC – "acceleration directive"; Directive 2003/55/EC of the European Parliament and of the Council of 26 June 2003 concerning common rules for the internal market in natural gas and repealing Directive 98/30/EC – "acceleration directive"; Christian Schütte, in: *Energiewirtschaftsgesetz,* ed. Martin Kment, (Baden-Baden: Nomos, 2015) para. 21 rec. 7; Uwe Pritzsche and Vivien Vacha, "Energierecht: Einführung und Grundlagen" (München: C.H. Beck, 2017), rec. 46.
4 Mohr, "Sicherung der Vertragsfreiheit durch Wettbewerbs- und Regulierungsrecht", p. 552.
5 Pritzsche and Vacha, "Energierecht: Einführung und Grundlagen", para. 2 rec. 47–48.
6 Directive 2009/72/EC of the European Parliament and of the Council of 13 July 2009 concerning common rules for the internal market in electricity and repealing Directive 2003/54/EC; Directive 2009/73/EC of the European Parliament and of the Council of 13 July 2009 concerning common rules for the internal market in natural gas and repealing Directive 2003/55/EC; Jens-Peter Schneider, in: *Recht der Energiewirtschaft,* ed. Jens-Peter Schneider and Christian Theobald (München: C.H. Beck, 2013) para. 2 rec. 46.

replaced by Regulation (EC) 714/2009 on the same day.[7] One of the main objectives of the Third Electricity Directive is the even greater separation of grid operation from production and supply (legal unbundling, operational unbundling, informational unbundling, accounting unbundling).[8] In chapter four, "Operation of the transmission grid", Articles 13 and 14, the new model of an "Independent system operator (ISO)" is intended to achieve the independent management of the grid company.[9] Chapter five, "Independent Transmission System Operator (ITO)", Articles 17-23, provides for an alternative to ownership unbundling and to the legal structure of the Independent System Operator.[10] Chapter eight, Articles 32-34, deals with the organisation of network access.[11] The aims of the directive are the reinforcement of the provisions concerning the separation of the transmission activity and the power of regulators, furthermore the promotion of an efficient use (with market mechanisms) of the infrastructures and construction of new interconnection.[12] The content of the provisions is consistent with Articles 20-22 of the Second Electricity Directive.[13] Their implementation was carried out by the provisions in Paragraph 20 to Paragraph 23a EnWG.[14] In addition, Article 14 of the Electricity Trading Regulation (EC) 714/2009 reinforces the model of an efficient and structurally comparable network operator as a benchmark for the pricing of cross-border electricity trading. The regulation intensifies the obligation of transmission system operators to cooperate with each other by establishing the ENTSO (Articles 4, 5) and its authority to establish network codes (Article 6).[15] The implementation of the directives in German

---

7 Regulation (EC) No 714/2009 of the European Parliament and of the Council of 13 July 2009 on conditions for access to the network for cross-border exchanges in electricity and repealing Regulation (EC) No 1228/2003; Pritzsche and Vacha, "Energierecht: Einführung und Grundlagen", para. 2 rec. 55.
8 Mohr, "Sicherung der Vertragsfreiheit durch Wettbewerbs- und Regulierungsrecht", p. 552; Jens-Peter Schneider, in: *Recht der Energiewirtschaft*, para. 2 rec. 48 et seq.
9 Cf. Pritzsche and Vacha, "Energierecht: Einführung und Grundlagen", para. 2 rec. 56.
10 Pritzsche and Vacha, "Energierecht: Einführung und Grundlagen", para. 2 rec. 56; Säcker, in: *Berliner Kommentar zum Energierecht*, Vol. 1, Einl. A. rec. 35.
11 Directive 2009/72; Directive 2009/73; Franz Jürgen Säcker, "Das Dritte Energiepaket für den Elektrizitätsbereich" (Frankfurt am Main: Peter Lang – Internationaler Verlag der Wissenschaften 2009), p. 7.
12 Directive 2009/72; the same objectives are pursued by the Directive 2009/73.
13 Directive 2003/54; cf. Art. 18-24 Directive 2003/55.
14 Säcker, "Das Dritte Energiepaket für den Elektrizitätsbereich", p. 7.
15 Regulation (EC) 714/2009; Säcker, "Das Dritte Energiepaket für den Elektrizitätsbereich", p. 8 et seq.

law led to the adoption of the Energy Industry Act (EnWG). The law came into force on 13 July 2005.[16] According to Paragraph 1 EnWG, a "safe, inexpensive, consumer-friendly, efficient and environmentally friendly supply of the general public with electricity and gas" is to be ensured. Regulation is intended to ensure "effective and undistorted competition". The fees for network access should be "reasonable, non-discriminatory and transparent", according to Paragraph 21 EnWG. Pursuant to Paragraph 1 Section 3, the purpose of the EnWG is also to implement and enforce European Community law in the field of grid-bound energy supply. The new provisions of Paragraph 20 et seq. EnWG ensure access to the entire electricity and gas supply network. With these provisions, the purpose of the regulation of grid-bound energy supply becomes clear: The opening of the networks to upstream and downstream competition.[17] Access regulation compensates for the fact that grid interconnection is a natural obstacle to competing energy supplies.[18]

2 Higher Regional Court of Düsseldorf

The Higher Regional Court of Düsseldorf is nationwide responsible for appeals against decisions of the German Federal Network Agency, the Bundesnetzagentur (BNetzA), in the field of grid-bound energy supply. In addition, the regulatory authorities of the sixteen federal states and their Higher Regional Courts are responsible for smaller grid operators. The Higher Regional Court of Düsseldorf has set up a senate solely for these proceedings. The 3rd Cartel Senate was established on the 1st of January 2006. Only a legal appeal to the Federal Court of Justice, the Bundesgerichtshof (BGH), is possible against its decisions. In order to judge economic issues, it is a necessity for the senate to employ economic experts. The judges have to fully understand all the economic methods applied by the German Federal Network Agency and being subject to judicial review. This is by nature time consuming, and therefore, the judges remain members of the senate for several years, so as to gain the necessary expertise. According to Paragraph 75 Section

---

16 Mohr, "Sicherung der Vertragsfreiheit durch Wettbewerbs- und Regulierungsrecht", p. 554.
17 Ibid., p. 579.
18 For more details on the natural monopoly as a reason for regulation: Mohr, "Sicherung der Vertragsfreiheit durch Wettbewerbs- und Regulierungsrecht", pp. 526 et seq.; about the importance of unbundling: Konstantina Bourazeri, "Verteilernetzentflechtung und Energiewende", *Recht der Energiewirtschaft* (2017), pp. 446 et seq.

2 of the Energy Industry Act (EnWG), the legal persons who were involved in the proceedings before the Federal Network Agency are in principle entitled to appeal against its decisions to the Higher Regional Court of Düsseldorf. In addition to the formal position as a party to the administrative proceedings, the complainant must also have a legal interest in legal protection.[19] Both a formal complaint and a substantive complaint are required. The latter condition is fulfilled, if the complainant claims a direct and individual impairment of his economic interests.[20]

3 The *Lichtblick* case

The complainant *Lichtblick* offers electricity and gas nationwide.[21] In July 2016, the Federal Network Agency initiated a procedure to determine the equity capital interest rates for electricity network operators for the third period of the German Incentive Regulation.[22] The complainant *Lichtblick* appealed against the determination of the equity capital interest rates, claiming the interest rates to be unreasonably high.[23] The Higher Regional Court of Düsseldorf rejected the complaint as inadmissible.[24] For the court, the complainant was not directly and individually affected by the fixing of the interest rates.[25] The fixing was only a preliminary question in the agreement of the network charges to be paid by the complainant, but the network operator had a certain margin by implementing the decision of the Federal Network Agency.[26] The Federal Court of Justice annulled, however, the decision of the OLG Düsseldorf and dismissed the proceedings.[27] It accepted the power of appeal and thus the admissibility of the complaint. For the Federal Court, the complainant had not been infringed

---

19 Anke Johanns and Katrin Roesen, in: *Berliner Kommentar zum Energierecht,* ed. Franz Jürgen Säcker (Frankfurt am Main: Fachmedien Recht und Wirtschaft, 2019), Vol. 1, para. 75 rec. 32 et seq.
20 Ibid., para. 75 rec. 36.
21 Higher Regional Court of Düsseldorf, judgment of 10 January 2018, VI-3 Kart 1202/16 (V), para. 1; German Federal Court of Justice, judgment of 9 July 2019, EnVR 5/18, para. 1.
22 Higher Regional Court of Düsseldorf, judgment of 10 January 2018, op. cit., para. 2; German Federal Court of Justice, judgment of 9 July 2019, cit., para. 1.
23 Higher Regional Court of Düsseldorf, judgment of 10 January 2018, cit., para. 4; German Federal Court of Justice, judgment of 9 July 2019, cit., para. 2.
24 Higher Regional Court of Düsseldorf, judgment of 10 January 2018, cit., para. 14.
25 Ibid., para. 23.
26 Ibid., para. 32.
27 German Federal Court of Justice, judgment of 9 July 2019, cit., para. 4.

in its own subjective rights, as the determination of the equity interest rates still required implementation by the network operators.[28]

The determination of the equity capital interest rates by the German Federal Network Agency, however, directly affected the economic interests of the complainant.[29] It is true that the determination addressed the network operators.[30] However, the determination is directly incorporated into the network operators' decision on the level of the network charges. The network operators, on the other hand, have no significant decision-making to do.[31] For business reasons, network operators have to take the fixed interest rate as a basis for the calculation of the network fees, which are the main profit source of network operators, in order to achieve the necessary return for the investors.[32] Therefore, the electricity supplier *Lichtblick* has been seen as materially directly and individually affected by the determination of the German Federal Network Agency.[33] The Higher Regional Court of Düsseldorf must now decide on the validity of *Lichtblick's* complaint.

So far, the net users, in this case, the current gas suppliers, could not proceed with their own complaints against the determination of the net fees by the Federal Network Agency. They had to appeal to the civil courts.[34] The consequence of the ruling by the Federal Court of Justice is that the court must now also decide on the appeal of the energy supplier as a grid customer and examine whether the level of equity interest rates is appropriate.[35]

4 Incentive regulation of the network operators – the X-factor

Network operators are subject to incentive regulation.[36] The current incentive regulatory formula goes as follows:

$$EO_t = KA_{dnb,t} + \left(KA_{vnb,t} + (1-V_t) \cdot KA_{b,t} + \frac{B0}{T}\right) \cdot \left(\frac{VPIt}{VPI0} - \mathbf{PF_t}\right) + KKA_t + Q_t + (VK_t - VK_0) + S_t$$

---

28 Ibid., para. 15.
29 Ibid., para. 18.
30 Ibid., para. 16, 21.
31 Ibid., para. 21 et seq.
32 Ibid., para. 23.
33 Ibid., para. 26.
34 Bernd Holznagel and Maximilian Hemmert-Halswick, "BGH: Entscheidung zur materiellen Beschwer von Netznutzern von der Festlegung des Eigenkapitalzinssatzes", *Zeitschrift für das gesamte Recht der Energiewirtschaft* (2019), 403, 407.
35 Ibid.
36 In detail Jörg Meinzenbach, "Die Anreizregulierung als Instrument zur Regulierung von Netznutzungsentgelten im neuen EnWG" (Baden-Baden: Nomos, 2008), pp. 151 et seq.

To this purpose, the costs of a base year are being recorded.[37] They form the starting level for determining the revenue ceilings for the subsequent 5-year regulatory period.[38] The revenue ceilings, i.e. the amounts which the grid operator may calculate annually as grid fees in the coming regulatory period,[39] are determined in accordance with this regulatory formula. The most important components are the costs that cannot be influenced permanently (KA_{dnb}), the costs that cannot be influenced temporarily (KA_{vnb}), and the costs that can be influenced (KA_b). The controllable costs are used to determine whether the *individual* company is efficient.[40] Existing inefficiencies must be reduced over the upcoming regulatory period.[41] Another component of the incentive regulatory formula is the so-called general sectoral productivity factor (– PF_t), also known as Xgen or X-factor. This, for me, is a highlight of the regulatory formula. Paragraph 9 Section 1 ARegV describes the X-factor in more detail: "The general sectoral productivity factor is determined according to the deviation of the network-specific productivity progress (ΔTF^{Netz}) from the macroeconomic productivity progress (ΔTF^{Ges}) and [the deviation] of the macroeconomic input price development ($\Delta Input^{Ges}$) from the network-specific input price development ($\Delta Input^{Netz}$)." Mathematically expressed, it means: $Xgen = \Delta TF^{Netz} - \Delta TF^{Ges} + \Delta Input^{Ges} - \Delta Input^{Netz}$

This general sectoral productivity factor has major economic importance for the network operators. The calculation of the X-factor is based on highly complex econometric methods. The German legislator considers the Törnqvist-Index and the Malmqvist-Index as suitable methods for calculating the X-factor.[42] These are internationally recognised econometric methods

---

37 Christian Theobald, Ines Zenke and Heiko Lange, in: *Recht der Energiewirtschaft*, eds. Jens-Peter Schneider and Christian Theobald (München: C.H. Beck, 2013) para. 17 rec. 54; Pritzsche and Vacha, "Energierecht: Einführung und Grundlagen", para. 4 rec. 307.
38 Cf. Para. 3 Sec. 2 ARegV; Pritzsche and Vacha, "Energierecht: Einführung und Grundlagen", para. 4 rec. 306 et seq.
39 Cf. Christian Theobald, Ines Zenke and Heiko Lange, in: *Recht der Energiewirtschaft*, para. 17 rec. 58 et seq.; Pritzsche and Vacha, "Energierecht: Einführung und Grundlagen", para. 4 rec. 320.
40 Pritzsche and Vacha, "Energierecht: Einführung und Grundlagen", para. 4 rec. 314.
41 Cf. Para. 16 Sec. 1 sentence 1 ARegV; Pritzsche and Vacha, "Energierecht: Einführung und Grundlagen", para. 4 rec. 314.
42 BR-Drucks. 417/07, pp. 48 et seq.; for the requirements on a transparent and comprehensible application of economic methods by the regulatory authorities see Mohr, "Anreizregulierung, Ökonomie, Mathematik - am Beispiel des generellen sektoralen Produktivitätsfaktors gem. § 9 ARegV", *Recht der Energiewirtschaft* (2020), 385, 389 et seq.

which measure productivity.[43] The X-factor is related to various legal questions which cannot be discussed in a narrow timeframe. The main principles are: The X-factor should reflect the development of productivity in the network industry in relation to macroeconomic productivity. Network operators should be able to observe the increase in network productivity and pass it on to their customers in the form of lower network charges.[44] The X-factor contains a general instrument for determining the revenue caps (EO_t). It "corrects" the inflation rate (VPI), which refers to the economy as a whole, by taking into account network-specific technological progress and network-specific prices.[45] The X-factor functions in the regulatory formula as a general offset (– **PF_t**) from the inflation rate (VPI).[46] In case of a positive development in productivity in the grid sector, the X-factor, thus, leads to a reduction of the allowed revenue caps.[47] The Higher Regional Court of Düsseldorf has to deal with a considerable amount of appeals of network operators against the determination of the X-factor by the German Federal Network Agency. The costs of the legal proceedings sum up to several billions of euros. In their complaints, network operators criticise the specific methods and the data used by the regulatory authority to calculate the X-factor. The Higher Regional Court of Düsseldorf has already announced its first decisions on these important issues on July 10, 2019. The Court has annulled the determination of the Federal Network Agency considering the X-factor in the gas sector.[48] The most important reasons for this decision are:

---

43 "Measuring Productivity – OECD Manual: Measurement of Aggregate and Industry-level Productivity Growth", rec. 154 et seq., OECD, July 16, 2001, http://www.oecd.org/sdd/productivity-stats/2352458.pdf.
44 BT-Drucks. 17/7632, p. 4; Andrea Berndt, "Die Anreizregulierung in den Netzwirtschaften" (Baden-Baden: Nomos, 2011), p. 179; Jörg Meinzenbach, "Die Anreizregulierung als Instrument zur Regulierung von Netznutzungsentgelten im neuen EnWG", p. 198.
45 Higher Regional Court of Düsseldorf, judgment of 10 July 2019, VI-3 Kart 721/18 (V), para. 2.
46 Jörg Meinzenbach, in: *Berliner Kommentar zum Energierecht,* ed. Franz Jürgen Säcker (Frankfurt am Main: Fachmedien Recht und Wirtschaft, 2019), Vol. 1, Para. 21a ARegV rec. 133.
47 Higher Regional Court of Düsseldorf, judgment of 10 July 2019, VI-3 Kart 721/18 (V), para. 3.
48 Higher Regional Court of Düsseldorf, judgment of 10 July 2019, VI-3 Kart 721/18 (V).

4.1 Relevant observation period (Stützintervall)

In the Court's view, the determination of the Xfactor by the Federal Network Agency is illegal because the result is not robust against changes in the observation period selected by the Federal Network Agency.[49] In the observation period 2006–2016, strong fluctuations become apparent, which the Federal Network Agency should have examined in detail for their causes.[50] An interval 2011–2016, for example, yields an X-factor of +2.95 %, an interval 2012–2016 a value of -2.12 %.[51]

4.2 Analogue application of the provisions about individual efficiency

The German Federal Network Agency has calculated the X-factor for the gas sector using the Törnqvist method. In order to review the plausibility of its results, it also used the Malmqvist method, which is therefore legally relevant.[52] The Malmqvist-Index applied by the Federal Network Agency is very similar to the methodologies of the Data Envelopment Analysis (DEA) and the Stochastic Frontier Analysis (SFA), which are used to measure and compare the *individual* efficiency of network operators.[53] The Federal Network Agency also relied for X-factor's calculation on the data used for measuring individual efficiency in the past three regulatory periods.[54] Since the Federal Network Agency uses the same methodologies and data basis to calculate the general productivity of network economy as it does in the context of individual efficiency comparison, it must also take the legal provisions related to the latter into consideration.[55] This means it has to apply the Best-of-four in an analogue way by determining the X-factor on the basis of the best result in favour of the grid operators.[56]

---

49 Ibid., para. 87 et seq.
50 Ibid.
51 Ibid., para. 88.
52 Ibid., para. 13; for the calculation of the X-factor by the German Federal Network Agency based on these methods see Bourazeri, "Rechtliche und ökonomische Grundlagen des generellen sektoralen Produktivitätsfaktors gemäß § 9 ARegV", *Netzwirtschaften & Recht* (2020), 194 et seq.
53 Ibid., para. 134, 136.
54 Ibid., para. 136.
55 Ibid., para. 136.
56 Ibid., para. 131 et seq.

5 Poland wins OPAL gas pipeline dispute before EU court

On 10 September 2019, the Court of First Instance of the European Union annulled a decision of the European Commission from 2016 that had allowed Gazprom to make full use of the capacity of the Ostsee Pipeline Anbindungsleitung (OPAL) natural gas pipeline.[57] The background to this decision is the Federal Network Agency's competence – in agreement with the Commission – to exempt a gas transit pipeline from regulation according to Paragraph 28a of the German Energy Industry Act (EnWG). Two unanimous decisions of the Federal Network Agency and the European Commission are therefore required. If one of the decisions is revoked, the exemption from regulation is not possible. In case T-883/16, the Republic of Poland, supported by the Republic of Latvia and the Republic of Lithuania, opposes the view of the European Commission, supported by the Federal Republic of Germany. Pursuant to Article 263 TFEU, the Republic of Poland seeks the annulment of the Commission's Decision,[58] amending the conditions for the exemption of the OPAL gas pipeline that were granted according to Directive 2003/55 from the access and tariff provisions of third parties. With regard to annulling the exemption of the gas pipeline from regulation provided for by a public law contract between OPAL and the German Federal Network Agency, the Polish State is simultaneously disputing the Federal Network Agency before the Higher Regional Court of Düsseldorf, with the participation of OPAL and Gazprom.

5.1 The facts

The Baltic Sea pipeline connection (OPAL) is the western connection of the Nord Stream 1 gas pipeline. The pipeline was commissioned in 2011 and has a capacity of around 36.5 billion cubic meters per year (m^3/year).[59] OPAL has a length of approximately 470 kilometres. The entry point is located in the Baltic Sea in Germany. The exit point is in the Czech Republic. In 2009 50 % of the pipelines were exempted from the rules of Directive 2003/55 on third

---

57 General Court, judgment of 10 September 2019, case T-883/16, ECLI:EU:T:2019:567, *Poland v. Commission*.
58 European Commission, Decision on review of the exemption of the Ostseepipeline-Anbindungsleitung from the requirements on third party access and tariff regulation granted under Directive 2003/55/EC, of 28 October 2016, C(2016) 6950.
59 General Court, judgment of 10 September 2019, case T-883/16, ECLI:EU:T:2019:567, *Poland v. Commission*, para. 9.

party access and tariff regulation.⁶⁰ Gazprom was only able to reserve further quantities if Gazprom put a gas volume of 3 billion cubic meters per year into circulation via this pipeline ("gas release programme").⁶¹ As Gazprom did not implement the gas release program, only 50 % of the OPAL pipeline's transport capacity was used. In 2016, the German Federal Network Agency notified the European Commission of its intention to amend certain provisions of the 2009 exemption at Gazprom's request on the basis of Directive 2009/73.⁶² The main purpose of the amendment was to replace the capacity limitation with an obligation for OPAL OGT to offer at least 50 % of its operated capacity at auction.⁶³ On 28 October 2016, the Commission approved the changes envisaged by the Federal Network Agency.⁶⁴ The German regulatory authority subsequently amended the exemption by concluding a public service contract.⁶⁵ Poland applied to the Court of First Instance of the European Union for annulment of the Commission's decision of 2016.⁶⁶ Poland argued that the Commission's decision infringes, inter alia, the principles of energy security and energy solidarity.⁶⁷ The granting of a new derogation to OPAL would jeopardize the security of gas supply in the European Union, in particular in Central Europe. Gazprom Group companies could now divert additional gas volumes to the EU market by making full use of the capacity of the North Stream 1 pipeline. If the demand for natural gas in Central Europe does not increase, it is going to affect the conditions of the other pipelines, mainly the Jamal and the Brotherhood pipelines,⁶⁸ for the use of transmission services, which would result in competition with OPAL.

In its judgment of 10 September 2019, the Court of First Instance stated that the principle of energy solidarity not only imposes mutual assistance

---

60 German Federal Network Agency, decision of 25 February 2009, BK7-08-009 and German Federal Network Agency, decision of 7 July 2009, BK7-08-009.
61 German Federal Network Agency, decision of 7 July 2009, BK7-08-009.
62 General Court, judgment of 10 September 2019, case T-883/16, ECLI:EU:T:2019:567, *Poland v. Commission*, para. 13.
63 Ibid., para. 14.
64 European Commission, Decision of 28 October 2016, C(2016) 6950.
65 General Court, judgment of 10 September 2019, case T-883/16, ECLI:EU:T:2019:567, *Poland v. Commission*, para. 18; Higher Regional Court of Düsseldorf, judgment of 27 July 2017, VI-3 Kart 1203/16 (V), para. 2.
66 General Court, judgment of 10 September 2019, case T-883/16, ECLI:EU:T:2019:567, *Poland v. Commission*, para. 19, 29.
67 Ibid., para. 51.
68 Ibid., para. 62.

obligations where a Member State is in an emergency situation regarding its gas supply. The European Union and the Member States must also make efforts to avoid measures that could harm the interests of the EU and other Member States in terms of security of supply.[69] The application of this principle does not mean that EU energy policy should not have any negative impact on one or more Member States' energy interests in general. However, the EU institutions and the Member States would have to take into account the interests of the European Union and the Member States.[70]

The Commission had to decide whether the amendment to the OPAL pipeline exemption proposed by the German authority could harm the energy interests of other Member States. The Court of First Instance criticises that the principle of solidarity was not dealt with in the Commission's Decision 2016.[71] It is not apparent for the court that the Commission examined that principle. Both the original decision and the decision in 2016 concern only the impact on the security of supply in the EU in general.[72] The Commission found that the additional capacity at the German-Czech border would benefit all regions accessible from that location and that the additional capacity would not lead to the complete replacement of the other transit routes.[73] However, the Commission should have examined the effects of the exemption on Poland's security of supply.[74] Furthermore, the Commission had not examined the medium-term consequences – in particular for Polish energy policy – of the transfer of part of the natural gas, which was previously transported by the Jamal and Brotherhood pipelines to the Nord Stream 1/OPAL transit route.[75] These effects must be weighed against the increased security of supply at EU level.[76] The Court of First Instance found that the decision of the Commission was issued in breach of the principle of energy solidarity.[77] Therefore the Court annulled the Commission's decision.

---

69 Ibid., para. 73.
70 Ibid., para. 72.
71 Ibid., para. 79.
72 Ibid., para. 81.
73 General Court, judgment of 10 September 2019, case T-883/16, ECLI:EU:T:2019:567, *Poland v. Commission*, para. 81.
74 Cf. ibid., para. 78, 81.
75 Ibid., para. 82.
76 Ibid.
77 Ibid., para. 83.

5.2 Further procedure

After the Court of First Instance had rejected the application of the State of Poland for a temporary injunction on 21 July 2017, the Higher Regional Court of Düsseldorf also rejected the urgent applications of the state-owned Polish companies on 11 October 2017.[78] The preliminary assessment had shown neither a formal nor substantive illegality of the public law contract.[79] In particular, it was not certain at the time of the assessment that the expansion of transport capacity on the OPAL would lead to a permanent relocation and constantly reduced capacity utilisation of the Jamal and the Brotherhood pipelines.[80] Following the conclusion of the main proceedings before the Court of First Instance, it is now probable that the main proceedings will also continue before the Higher Regional Court of Düsseldorf.

---

78 Higher Regional Court of Düsseldorf, judgment of 11 October 2017, VI-3 Kart 1203/16 (V).
79 Ibid., see section B.II.1.
80 Ibid., see section B.II.1.2.1.

Theodoros Galanis[*]

4 Multilateral Investment Court – an alternative to energy disputes

Abstract: This chapter focuses on the new multilateral approach towards dispute settlement advocated by the EU. The EU is the world's biggest recipient and source of Foreign Direct Investment (FDI) and one of the most important global energy actors. Investments, including in the energy sector, can generate disputes between the investors and the host states. The current model of the ad hoc dispute settlement (ISDS) found in the vast majority of the international investment agreements does not sufficiently guarantee impartiality and predictability. The EU has therefore decided to move away from this old-style system and promote the creation of a multilateral investment court to rule on investment disputes. This multilateral approach could offer a new perspective to energy-related investment disputes.

Keywords: multilateral approach, investment court, institutionalized court system (ICS), energy disputes, criticism to ISDS

Introduction

The EU is the world's largest source and destination of Foreign Direct Investment (FDI), and, therefore, has a strong interest to facilitate and protect international investment and support European investors. As Trade Commissioner Cecilia Malmström underlined: "As the EU upholds a high standard in promoting and protecting investment in its territory, the EU has a natural interest in obtaining similarly credible and enforceable guarantees for EU investments and investors abroad".[1] The EU is committed to multilateralism and is one of the most vocal supporters of free trade and investment. The rationale underpinning this approach is that multilateralism creates an environment where trade and investment can prosper, and that trade and investment have been a powerful tool for growth and development when used in the right way. For trade and investment

---

[*] Teaching assistant – University of Piraeus. The opinions expressed in the paper are purely personal.
[1] Cecilia Malmström, "A Multilateral Investment Court: a contribution to the conversation about reform of investment dispute settlement", Speech of 22 November 2018, http://trade.ec.europa.eu/doclib/docs/2018/november/tradoc_157512.pdf.

to support a sustainable development, rules are needed, which should also be backed up with effective enforcement. Neutral and effective dispute settlement mechanisms are instrumental in achieving this. The EU Member States are parties to almost half of the total number of international investment agreements that are currently in force worldwide. These agreements, almost all of which include both investment protection and investor-to-State dispute settlement ("ISDS") have played their part in encouraging and protecting the high volume of EU investment abroad and, reciprocally, the investments held by the rest of the world in the EU.[2]

The ISDS system has worked for decades in a –more or less – acceptable way. However, recent developments fuelled doubts about its efficiency and fairness and point towards its reform: Firstly, the EU is actively seeking to overhaul investment dispute settlement. Unlike the WTO, the dispute settlement mechanisms for investment were not designed for their current purpose. They are outdated since they were introduced in the 1960s and are based on commercial arbitration, i.e. on a system for resolving private commercial disputes, not for the settlement of disputes involving States. This system has proven to be inadequate and exposed to criticism from the civil society who voiced concerns that ISDS does not sufficiently guarantee impartiality and predictability. Therefore, the EU has actively sought to change it, by including a new investment court system in its latest agreements with Canada, Singapore, Vietnam, Mexico, while also putting forward the idea to create a multilateral investment court. Secondly, the Lisbon Treaty conferred competence for the protection of investments to the EU. The Treaty gave the EU exclusive competence over foreign direct investment as part of its common commercial (trade) policy, alongside goods, services and trade-related aspects of intellectual property. EU trade policy developed to include negotiations for ISDS as part of trade negotiations, thus creating an unprecedented opportunity not only for a comprehensive approach to trade and investment at EU level, but also for a profound reform of the traditional approach to investment protection and the associated ISDS system. In addition, the multilateral system is currently under threat. In particular, the World Trade Organization (WTO), the body that has underpinned global trade for decades, is now being side-lined.

The EU has put forward a broad set of ideas to modernize WTO, including changes to the dispute settlement system and the strengthening of its processes.

---

2   ISDS allows to settle disputes between an investor and a State when the latter is alleged to have breached its commitments under an international investment agreement.

All the above parallel developments contribute to the creation of a political and legal environment conducive to an overhaul of the international dispute settlement system, including on energy cases. The energy sector has an inherent international and cross-border nature. Energy disputes have a particularly high share in the total number of claims for dispute resolution: According to the statistics of the International Centre for Settlement of Investment Disputes (ICSID), one of the most prominent international arbitration institutions, from 1972 to 2012 energy-related disputes accounted for more than 37 % of the claims registered under the ICSID Convention.[3] In light of the foregoing, the present chapter will focus on the new multilateral approach towards dispute settlement advocated by the EU. The chapter will first summarize the main criticism addressed to the old ISDS system, before presenting the highlights of the new EU approach, as enshrined in the recent CETA agreement. Subsequently, it will outline the basic features of the Multilateral Investment Court (MIC), which will be in the epicentre of the revamped dispute settlement system, and draw the attention to some concerns voiced on the compatibility of the new approach and the MIC with EU law. The chapter will conclude with the answers provided recently by the Court to the above questions, through the opinion 1/17.

1 Main criticism addressed to ISDS

The criticism levelled against the long-established ISDS, which also prompted the initiative of the European Commission (EC) to champion the reform of the current dispute settlement system, is concentrated on several critical points. These include the perceived lack of procedural transparency and consistency in the interpretation of legal issues, excessive costs of the proceedings, flaws related to the selection process and the impartiality of arbitrators, also raising serious legitimacy issues. Several stakeholders around the world share these concerns, also mirrored on the discussions speared by the United Nations Commission on International Trade Law (UNCITRAL), which started working on a possible ISDS reform at the end of 2017. UNCITRAL Working Group III, with the participation of the EU, is engaged in identifying potential areas for reform.[4] The

---

3   Andrey Konoplyanik, "Multilateral and bilateral energy investment treaties: Do we need a global solution? The Energy Charter Treaty as an objective result of the evolution of international energy markets and instruments of investment protection and stimulation", in: *Research Handbook on International Energy Law*, ed. Kim Talus (Cheltenham/Northhampton: Edward Elgar Publishing, 2014), pp. 79–123.
4   Pohl Joachim, Mashigo Kekeletso and Nohen Alexis, "Dispute settlement provisions in international investment agreements: A large sample survey", *OECD, International*

criticism embraced *inter alia* several issues. A first issue relates to the costs of the proceedings. The arbitration costs are quite high, especially for SMEs who are thus discouraged from engaging in arbitration proceedings and consider that their right to access justice is impaired. States are also allegedly overburdened and discriminated against, in the sense that when they lose a claim, damages tend to be very high. High arbitration costs also entail the involvement of third parties to the dispute, who can fund claims in exchange for a share of the potential final compensation awarded. It has been reported that investment companies (banks, hedge funds and insurance companies) are increasingly engaged into this 'third party funding of claims' phenomenon, which can result in an increase of frivolous cases against States.[5] Another issue was the duration of proceedings. The arbitration proceedings are quite lengthy, thus undermining legal certainty and neutralizing one of the primary arguments against recourse to normal judicial proceedings. An additional issue relates to concerns over the lack of transparency and legitimacy. Confidentiality is a crucial feature of arbitration proceedings. Confidentiality caveats hinder access to crucial details of the cases and lead to allegations about undemocratic processes and undue policy-making by large multinational companies and strong vested interests. Non-governmental public interest groups and representatives from civil society have traditionally been excluded from the proceedings.

Fear of potential reduction in states' sovereign power to regulate (the so-called chilling effect) was one more issue of criticism. States prefer not to regulate their public interests in order to avoid any liability, especially under rules on indirect expropriation and legitimate expectations. Another issue relates to allegations about forum shopping from investors, who either initiate parallel proceedings in domestic courts and through ISDS or choose the arbitration forum that is most likely to provide a favourable judgment. Lack of independence, impartiality, and neutrality of adjudicators in practice, has been one more issue of criticism. The parties of the disputes select the members of the arbitration tribunals who do not receive any permanent remuneration. Although all arbitration fora require that arbitrators exhibit independence and impartiality in

---

Investment Working Paper, no. 2, November 2012, https://papers.ssrn.com/sol3/papers.cfm?abstract_id=2187254.

5   Marta Latek and Laura Puccio, "Investor – State Dispute Settlement (ISDS) State of play and prospects for reform", *European Parliamentary Research Service*, January 2015, p. 5, http://www.europarl.europa.eu/thinktank/en/document.html?reference=EPRS_BRI(2015)545736

the handling of cases, there are no solid indications that their appointment and working methods indeed meet in practice such requirements.[6] It appears that arbitrators, who come from a relatively closed specialized circle, are not often bound by strict conflict of interest and impartiality obligations. The arbitrators' pool is rather restricted, and it is not uncommon to witness a change of roles, whereby an investment lawyer is successively, e.g. a counsel to a firm, arbitrator and expert in different cases. Finally, one more issue of criticism relates to the absence of appeal mechanisms. "There are few ways to review the decision of an arbitral tribunal since most tribunals are *ad hoc*, deciding a case on the merits, without the possibility to launch an appeal against their decision. Awards can be set aside, but on limited grounds that do not include errors of law or fact."[7] The absence of an appeal mechanism aggravates the risk of divergent interpretations of similar provisions and undermines legal certainty and predictability of the awards. On the other hand, defenders of the current system argue that the above concerns are exaggerated and ISDS has delivered, achieving its main objective to attract investments while securing property rights. The set-up of ISDS has the objective – which has been largely met – to reduce the political risks inherent in rapidly increasing foreign investments and allow more effective enforcement of host investment countries' commitments. Within the context of this politicized debate, the European Union has taken the lead in the ISDS reform process.

2 The new multilateral approach

Based on its new competences in the field of investment policy, as well as on its overall economic weight and the vast number of bilateral investment treaties (BITs) concluded by the Member States, the EU has been a key player in discussions for the strengthening of a rule-based multilateral system, through the reform of the traditional approach to investment protection and the associated ISDS system. After all, multilateral instruments have several comparative benefits against bilateral ones, since they provide a common denominator,

---

6   Jane Kelsey, "'Investor-State' Disputes in Trade Pacts Threaten Fundamental Principles of National Judicial Systems", *Public Citizen,* April 2012, https://www.citizen.org/wp-content/uploads/migration/isds-domestic-legal-process-background-brief.pdf.
7   Erika Szyszczak, "Opinion 1/17: Towards a Modern EU Approach to Investor-State Dispute Settlement", *UK Trade Policy Observatory, Briefing Paper 30,* May 2019, http://blogs.sussex.ac.uk/uktpo/files/2019/05/BP-30-interactive1.pdf.

thus avoiding extra costs, saving time and minimising efforts for their negotiation.[8] The EU itself, through the Energy Charter Treaty, and its Member States are parties in around 1400 agreements, which provide for ISDS.[9] The entry into force of the Lisbon Treaty gave a new impetus to the EU work in this field. In terms of publicity though, the issue of investment protection and the related ISDS mechanism started attracting significant public interest during the negotiations on a Comprehensive Economic and Trade Agreement (CETA) with Canada and a Transatlantic Trade and Investment Partnership Agreement (TTIP) with the United States. In 2014, the EU concluded the negotiations for the first two free-trade agreements that include investment protection and ISDS, with Canada (CETA) and Singapore (EU-Singapore Free Trade Agreement). These agreements introduce a new approach, both in terms of substance (investment protection rules) and procedure (ISDS mechanism).

Public interest on these issues culminated during TTIP negotiations that triggered reactions from civil society organizations, with much of the attention focusing on the provisions on investment protection and ISDS. In this context, the European Commission launched a public consultation in order to gather views from the public on how the EU could develop further its approach on investment protection and ISDS. The consultation has allowed the identification of four main areas for further improvement: (1) the protection of the right to regulate; (2) the supervision and functioning of arbitral tribunals; (3) the relationship between ISDS arbitration and domestic remedies, and (4) the review of ISDS decisions for legal correctness through an appellate mechanism. The results of the consultation were followed by a concept paper in May 2015, which sums up the approach of the Commission, towards achieving greater transparency, consolidating the "right to regulate" and restricting controversial provisions of international investment law.[10] With regard to the reform of the traditional ISDS system, the discussion paper sets out a two-step approach: Initially, the inclusion of an institutionalized court system (ICS) for the resolution of investment disputes in all future EU trade and investment agreements with other negotiating partners. Subsequently, the establishment of a multilateral investment court, which would ultimately replace all the bilateral

---

8 Andrey Konoplyanik, p. 79.
9 Latek Matek and Puccio Laura.
10 "Investment in TTIP and beyond – the path for reform – Enhancing the right to regulate and moving from current ad hoc arbitration towards an Investment Court", *EC Concept Paper,* 5 May 2015, https://trade.ec.europa.eu/doclib/docs/2015/may/tradoc_153408.PDF.

ICSs included in the Union trade and investment agreements. The creation of a system where adjudicators are appointed by the Contracting Parties (and not by the parties to the dispute) would already move ISDS procedures closer to a permanent court. ISDS would then be institutionalized even further by establishing an actual permanent investment court with tenured judges. This court would apply to multiple agreements and between different trading partners, also on the basis of an opt-in system. According to the paper, "the objective would be to multilateralise the court either as a self-standing international body or by embedding it into an existing multilateral organization".[11] As already indicated, CETA was the first agreement signed by the EU to include provisions for the protection of investments and the establishment of a dispute settlement mechanism to enforce these protection rights. The new investment court system that was introduced in the agreement is to serve as a model for other trade negotiations that the EU is either conducting or will be engaged into in the future. The negotiations on a free trade deal between the European Commission and the Canadian Government were initially concluded in 2014 with a reformed investment dispute settlement system. However, the legal revision of the text pushed the reform even further; the final text contains all the main elements of the EU's new approach on investment which were outlined in the EU proposal for TTIP (Nov. 2015) and were included in the subsequently concluded EU-Vietnam free trade agreement. The revised text represents a clear break from the traditional ISDS and signals the EU's intent to present this new proposal on investment in its future negotiations with all partners. In particular, stronger language has been introduced on the right to regulate regarding investment protection. The parties preserve their right to regulate and to achieve legitimate policy objectives, such as public health, safety, environment, public morals and the promotion and protection of cultural diversity.[12]

Furthermore, key concepts like "fair and equitable treatment" and "indirect expropriation" have been defined for the first time in order to prevent abuse.[13] In respect of dispute settlement, a permanent and institutionalized dispute settlement tribunal, whose members will be appointed in advance by the Parties of the agreement and no longer by the investor and the State involved in the dispute, will replace the ad hoc arbitration system. Further to that, concrete measures have been taken, firstly to prevent forum shopping, including the

---

11 Ibid., pp. 11–12.
12 Article 8.9 of the Agreement.
13 Articles 8.10 and 8.12 of the Agreement.

prohibition of "mailbox companies" bringing cases to arbitration and secondly to ensure early dismissal of unfounded claims: a fast-track system will allow the early rejection of manifestly unfounded or frivolous claims.[14] Frivolous claims will also be discouraged by the introduction of the "loser pays" principle, which will also entail that the investor will pay the litigation costs of the state he has challenged.[15] In addition, full, mandatory transparency of the arbitration process has been introduced on the basis of the UNCITRAL rules, while parallel proceedings are not allowed: investors cannot seek remedies in domestic courts and through ISDS at the same time.[16] This way, divergent interpretations of the provisions will be significantly reduced. As regards the interpretation of the rules, it should also be noted that governments – not arbitrators – will have the ultimate control. Under CETA, the EU and Canada can issue binding interpretations on specific rules, and the ISDS Tribunal is obliged to follow these interpretations. The revised CETA stipulates that the Tribunal shall only apply the agreement, in accordance with the principles of international law, when adjudicating upon claims submitted by investors. It cannot decide on matters of EU or Member State law but can only look at EU or Member State law as a matter of fact.[17] It will therefore not interpret EU or Member States law in a manner binding on EU courts or EU governments. The establishment of a Tribunal, as already highlighted, as well as of an Appellate Tribunal, is perhaps the most innovative feature of the agreement.[18] Inspiration has been drawn from the quasi-judicial system of dispute settlement created for the WTO Appellate Body.[19] The permanent CETA Tribunal, which will be competent to decide upon investment disputes and claims for violation of the investment protection standards set by the Agreement, will be comprised of 15 members.[20] The

---

14 Article 8.32 of the Agreement.
15 At present, even if a government successfully defends itself, it often has to bear its litigation costs.
16 Article 8.22 of the Agreement.
17 For example, Member State law would be applied in order to make sure that the property rights in question are in fact held by the investor.
18 Articles 8.27 (Tribunal) and 8.28 (Appellate Tribunal) of the Agreement.
19 Laura Puccio and Roderick Harte, "From arbitration to the investment court system (ICS): The evolution of CETA rules", *European Parliamentary Research Service*, June 2017, http://www.europarl.europa.eu/RegData/etudes/IDAN/2017/607251/EPRS_IDA(2017)607251_EN.pdf.
20 It is also worth mentioning that because of the expected low number of cases, CETA does not provide for the creation of new facilities but instead foresees the use of

Judges will be appointed for a pre-defined term by the CETA Joint Committee and not by the disputing parties, and a three-member division of the Tribunal will be responsible for the hearing of each particular case. The composition of the Appellate Tribunal has not yet been defined. The members of both the Tribunal and the Appellate Tribunal will be bound by a detailed code of conduct, guaranteeing the respect of high ethical and professional standards. The Agreement includes specific provisions on the procedures, which must be followed to ensure the impartiality of arbitrators and the concrete steps to be taken in order to determine whether a conflict of interest could arise or has arisen. For instance, members of the Tribunal are not allowed to work as lawyers or experts in any other investment disputes.

The same restrictions apply to the Members of the Appellate Tribunal; CETA is the first agreement to include an Appeal mechanism, which will allow the correction of errors and ensure the consistency of the decisions of the Tribunal of first instance. An appeal can be launched within 90 days of the final awards, allowing for a full review of the case.[21] The bilateral mechanism described above is only an intermediate step towards the establishment of a new permanent multilateral dispute resolution system. The EU and the Canadian Government committed to join forces with other trading partners and to step up their efforts for the setting up of a permanent multilateral investment court (MIC).[22] The text of CETA recognizes that the MIC, which will be examined in more detail below, will eventually replace the bilateral mechanism established in CETA. To sum up, CETA represents a radical change in investment rules and dispute resolution; it moves away from the traditional approach of investment dispute resolution by promoting the creation of independent, permanent investment tribunals. Further to that, the agreement lays down the foundations for transforming this new approach into a multilateral investment court, with both parties committing to work expeditiously into this direction. The objective is to attract a critical mass of participants to the venture that would replace bilateral systems, including CETA, by the new court, which would be open to accession by any country.[23]

ICSID as an administrative secretariat that supports the ICS proceedings from an organizational and logistical point of view.
21 Point 8.28 of the Agreement.
22 Ibid., point 8.29.
23 Joint Interpretative Instrument on CETA, OJ L 11, 14 January 2017, p. 3, para. 6. At the time of signature of the CETA, the European Union and its Member States and Canada established this Instrument.

3 The Multilateral Investment Court

The reference to the MIC in CETA has by no means taken by surprise the business community and the academic commentators. Replacing the outdated ISDS system was one of the pledges made by President Juncker in his October 2014 speech in front of the European Parliament.[24] The EU's intention to move towards this direction is also reflected on the Commission Communication of 2015 "Trade for all" according to which the Commission will – in parallel to its bilateral efforts – "engage with partners to build consensus for a fully-fledged, permanent International Investment Court".[25] Likewise, the European Commission's statement at the occasion of the release of the EU's proposed text for TTIP stresses the starting of the work "together with other countries, on setting up a permanent International Investment Court".[26] Similar references can also be traced – apart from the 2015 concept paper already presented – in the 2017 Commission Reflection Paper on Harnessing Globalization and in Council conclusions.[27] The idea is to establish an international court with judges nominated for set terms –a *sine qua non* for their independence – empowered to hear disputes pertaining to investments between investors and States who have accepted its jurisdiction over their BITs. The MIC should ideally be for investment dispute settlement what the WTO is for trade dispute settlement. As for the reasons behind the EU choice, the Trade Commissioner Cecilia Malmström epitomizes the rationale underpinning the reform of the traditional investment arbitration model, leading to the creation of MIC.

---

24 Jean-Claude Juncker, "Time for Action – Statement in the European Parliament plenary session ahead of the vote on the College", Speech of 22 October 2014, https://ec.europa.eu/commission/presscorner/detail/en/SPEECH_14_1525.
25 European Commission, Trade for All: Towards a more responsible trade and investment policy, COM/2015/0497 final of 14 October 2015, https://trade.ec.europa.eu/doclib/docs/2015/october/tradoc_153846.pdf.
26 European Commission, EU Finalises Proposal for Investment Protection and Court System for TTIP, Press release IP/15/6059, 12 November 2015, http://europa.eu/rapid/press-release_IP-15-6059_en.html.
27 European Commission, A Balanced and Progressive Trade Policy to Harness Globalisation, COM(2017) 492 final of 13 September 2017, https://ec.europa.eu/commission/sites/beta-political/files/reflection-paper-globalisation_en.pdf, ("[international investment] [d]*isputes should no longer be decided by arbitrators under the so-called investor-state dispute settlement. This is why the Commission has proposed a multilateral investment court that would create a fair and transparent mechanism*").

Effectiveness and independence–impartiality are the keywords: As the Commissioner points out, only a permanent body with full-time adjudicators, fairly distributed in geographical and gender terms, supported by an effective appeal mechanism can ensure predictability and consistency of the decisions, while eliminating ethical concerns.[28] The costs and the duration of the proceedings will also be significantly curtailed, thus facilitating SMEs' access to the dispute resolution system. It should be recalled that the EU is the world's largest exporter and importer of FDI and therefore has a strong interest in ensuring that resolution of disputes operates smoothly. Canada and numerous other governments and international organizations share the same concerns. In this context, UNCITRAL agreed to discuss the ISDS reform in 2017 and in March 2018, the Council mandated the Commission to start negotiations with its trading and investment partners in the framework of UNCITRAL.[29]

4 Basic features of MIC – the Council mandate

Pursuant to Article 218 TFEU on the conclusion of international agreements, the Council authorizes the opening of negotiations, adopts negotiating directives, authorizes the signing of agreements and concludes them. On this legal basis, the Council adopted in March 2018 negotiating directives authorizing the Commission to negotiate a convention establishing a multilateral court for the settlement of investment disputes.[30] The MIC envisioned by the European Union would be a permanent international institution to adjudicate disputes under both existing and future investment treaties. While the exact features of the MIC will, naturally, depend on the outcome of the on-going negotiations under the auspices of UNCITRAL, the general principles favoured by the EU and included in the Council mandate are the following: In organizational terms, the court should be a permanent international institution, composed

---

28   Speech by European Commissioner for Trade Cecilia Malmström, op. cit.
29   European Commission, Recommendation for a Council Decision: Authorizing the opening of negotiations for a Convention establishing a multilateral court for the settlement of investment disputes, Commission Staff Working Document, Impact Assessment, Multilateral reform of investment dispute resolution, SWD(2017) 302 final of 13 September 2017.
30   Council of the European Union, Negotiating directives for a Convention establishing a multilateral court for the settlement of investment disputes of 20 May 2018, http://data.consilium.europa.eu/doc/document/ST-12981-2017-ADD-1-DCL-1/en/pdf.

of a tribunal of first instance and an appeal tribunal. Its composition should guarantee impartiality and independence, which in turn require that the judges are tenured, qualified, receive permanent remuneration and adhere to strict ethical standards. The proceedings before the court should be conducted in a transparent manner, including the possibility of third parties to submit interventions. According to the Council's mandate, the effective enforcement of the court's decisions would be of paramount importance and should thus be ensured.

Similarly, the possibility of appeal against the decisions of the tribunal of first instance should be provided for, the appeal Tribunal being competent to review the decisions on the grounds of errors of law or manifest errors in the appreciation of facts or serious procedural shortcomings. The court should have jurisdictions over disputes under future and existing investment treaties that countries decide to assign to the authority thereof. Special attention is also given to the costs and the duration of the proceedings: the cost- and time-effective operation of the Court is an objective of the negotiations when reading the Council's mandate. The contracting parties should bear the fixed costs of the MIC with a possible contribution from the disputing parties through court fees, which should take into account – *inter alia* – the level of economic development of the parties and will not be linked to the remuneration of the Members of the court. According to the mandate given by the Council, the Convention establishing MIC "should be open for signature and accession by any interested country and regional economic integration organization that is a party to an investment agreement". The Council clarifies that intra EU-BITs and the intra-EU application of the Energy Charter Treaty (ECT) are not within the scope of the negotiating directives.[31]

5 Critical issues – some reflections

Concerns were raised about the compatibility of ICS and MIC with the principle of autonomy of the EU legal order. Opinion 1/09 on the European Patent Court and 2/13 on ECHR defended the European Court of Justice's (ECJ) exclusive power to give authentic interpretations of EU law to ensure its uniform interpretation across the EU. Consequently, an ICS system envisaging the creation of a Tribunal outside the EU legal system and the subsequent setting of MIC should preclude that such bodies issue interpretations of EU law that are

---

31 Footnote 1 of the conclusions.

binding on EU institutions. The direct link between the ECJ and the national courts is instrumental in this respect. According to the ECJ's opinion on the European Patent Court: "Consequently, the envisaged agreement, by conferring on an international court which is outside the institutional and judicial framework of the European Union an exclusive jurisdiction to hear a significant number of actions brought by individuals in the field of the Community patent and to interpret and apply European Union law in that field, would deprive courts of Member States of their powers in relation to the interpretation and application of European Union law and the Court of its powers to reply, by preliminary ruling, to questions referred by those courts and, consequently, would alter the essential character of the powers which the Treaties confer on the institutions of the European Union and the Member States and which are indispensable to the preservation of the very nature of European Union law".[32] Likewise, the agreement on the EU accession to ECHR was found incompatible with EU law, *inter alia* on grounds related to the respect of the preliminary rulings mechanism (Article 267 TFEU).[33]

Further to that, with its landmark *Achmea* case, the European Court of Justice found that the arbitration mechanism in EU internal bilateral investment treaties is incompatible with European law. The ruling makes it clear that investment arbitration between two member states (in this case, the Czech Republic and the Netherlands) is illegal and must be dismantled because it endangers the autonomy of the EU legal order. A three-pronged test was conducted by the Court in order to assess the compatibility of the respective BIT and in particular its ISDS system with EU law: An arbitral tribunal is compatible with EU law if, alternatively either, a) does not interpret EU law, or b) belongs to the judicial system of the Member States or, c) its awards are reviewable by the Member States' courts. The absence of these characteristics has adverse effects

---

32 European Court of Justice, opinion 1/09 of 8 March 2011, ECLI:EU:C:2011:1230, para. 89.
33 European Court of Justice, opinion 2/13 of 18 December 2014, ECLI:EU:C:2014:2454, *Accession of the European Union to the European Convention for the Protection of Human Rights (ECHR) and Fundamental Freedoms – Compatibility of the draft agreement with the EU and FEU Treaties*; further to the issues raised with regard to preliminary rulings, the Court did not rule out that disputes between Member States or between Member States and the EU concerning the application of the ECHR within the scope *ratione materiae* of EU law could be brought before the ECtHR, thus affecting Article 344 TFEU.

on the autonomy of EU law.[34] In the light of the foregoing, it appears that the Court considers the non-interference of the ISDS-ICS with the allocation of competences between the EU and the Member States, the respect of the preliminary ruling mechanism and the existence of some degree of control by the ECJ in case there is disrespect of EU law as important benchmarks for safeguarding the autonomy of the EU legal order. In this respect, the Commission and several Member States claimed that the approach adopted in the *Achmea* case could not be simply ruled over to the ICS or the MIC in free trade agreements. Indeed, in its opinion 1/17 on CETA at the Belgian government's request, which was submitted on the basis of Article 218 TFEU, the ECJ distinguished the CETA context from the intra-EU investment agreement at stake under *Achmea*.

The specifics of the opinion, which is particularly relevant for the establishment of MIC, will be exposed below. What should also be highlighted here and added to the EU-law related issues is the concern voiced by some stakeholders that the perceived threat to the Member States' right to regulate would continue to exist under a MIC as companies could still claim massive compensation from states in arbitral proceedings. In their view, MIC focuses on procedural issues, such as transparency, impartiality and accountability of judges and does not address substantial issues and provisions in international trade and investment agreements, such as the broad rights granted to investors, compliance with human rights, labour and environmental laws. Their main position is that the EU should focus on ensuring that States have the necessary policy space to develop new regulations for the protection of public interest and promote sustainable development.[35] In this regard, the position of the EU has been that a permanent court, composed of highly qualified judges whose independence is beyond doubt, should ensure that the right balance between the investors' rights and societal interest is found whenever the investment protection rules of existing treaties are interpreted.

5.1 Critical issues – some answers

The ECJ's opinion on the legality of the new bilateral ICS in CETA sheds light on most of the EU law-related concerns. In a nutshell, the Court examined the compatibility of the CETA investment provisions with EU law from three

---

34 European Court of Justice (Grand Chamber), judgment of 6 March 2018, case C-284/16, *Achmea*, para. 39–56.
35 Latek and Puccio, "Investor – State Dispute Settlement (ISDS) State of play and prospects for reform", op. cit., p. 8.

different angles: the autonomy of the EU legal order, equal treatment and effectiveness and the right of access to an independent tribunal. For each point, the Court, who – unlike the *Achmea* case – sided with the opinion of the Advocate General in this case, sets out the principles, before applying them to CETA. On the first and most important point, the Court recalled at the outset that "an international agreement providing for the creation of a court responsible for the interpretation of its provisions and whose decisions are binding on the European Union, is, in principle, compatible with EU law", provided there is no adverse effect on the autonomy of the EU legal order. It follows that "CETA, in so far as it provides [...] a process of submitting to judicial adjudication the resolution of disputes between investors and States by means of establishing a CETA Tribunal and Appellate Tribunal and, in the longer term, a multilateral investment Tribunal, may be compatible with EU law only if it has no adverse effect on the autonomy of the EU legal order".[36]

In order to determine the compatibility of the envisaged ISDS mechanism with the autonomy of the EU legal order, it is necessary to be satisfied that CETA "a) does not confer on the envisaged tribunals any power to interpret or apply EU law other than the power to interpret and apply the provisions of that agreement having regard to the rules and principles of international law applicable between the Parties and b) does not structure the powers of those tribunals in such a way that, while not themselves engaging in the interpretation or application of rules of EU law other than those of that agreement, they may issue awards which have the effect of preventing the EU institutions from operating in accordance with the EU constitutional framework".[37] In other words, the creation of courts outside the EU judicial system, like the CETA one, does not breach as such the autonomy of the EU legal order, if certain limits are respected. The CETA Tribunals "cannot have the power to interpret or apply provisions of EU law other than those of the CETA or to make awards that might have the effect of preventing the EU institutions from operating in accordance with the EU constitutional framework". They cannot, therefore, impact, on the one hand, directly on EU law, by interpreting or applying it and on the other hand indirectly, by issuing "awards which have the effect of preventing the EU institutions from operating in accordance with the EU constitutional framework". The Court decided that CETA's ICS does not affect EU

---

36 European Court of Justice, opinion 1/17 of 30 April 2019, ECLI:EU:C:2019:341, *CETA*, para. 108.
37 Ibid., para. 118.

law directly, since the power of interpretation and application conferred on the CETA Tribunal is confined to the provisions of the CETA, with no possibility for a preliminary reference to the ECJ. In addition, the CETA Tribunal does not indirectly impact the EU law through an indirect effect on the operation of the institutions, as it contains several provisions ensuring public interest considerations and the right of the Parties to regulate.[38] The Court then turned to the general principles of equal treatment and effectiveness and the right to access an independent Tribunal, only to find that the provisions of CETA did also not breach them. The detailed line of argumentation followed by the Court on these points is beyond the scope of this chapter. It should only be stressed that overall, the Parties were cautious in drafting the agreement and defining the jurisdiction of the CETA tribunal in order not to impinge on ECJ's competences.[39] Therefore, the Court was pleased that the autonomy of the EU legal order, the general principles and rights guaranteed by EU law, would not be adversely affected. It has, therefore, confirmed the compatibility of the agreement with EU law. The decision by the Court means that no changes to the text of the EU-Canada agreement are necessary and Member States' ratifications can proceed. Equally, no change will be required in the ICS provisions included in the agreements with Singapore, Mexico and Vietnam.

Now that most challenges related to MIC's compatibility with EU law have been addressed, it remains to be seen how much international support the EU can gather for this venture, especially from its major trading partners. Support from civil society should also be secured, as opponents of ISDS mechanisms argue that the establishment of a MIC will not alleviate concerns about investment arbitration. They maintain, *inter alia*, that the perceived threat to states' right to regulate would continue to exist under a MIC as companies could still claim massive compensation from states in arbitral proceedings.[40]

---

38 Ibid., para. 150.
39 Panos Koutrakos, "More on Autonomy – Opinion 1/17 (CETA)", *European Law Review* (2019), 3, 293–294, https://dialnet.unirioja.es/servlet/articulo?codigo=6976614.
40 See "Legislative Train 05.2019. 6A A Balanced and progressive trade policy to harness globalisation", May 2019, available at http://www.europarl.europa.eu/legislative-train/api/stages/report/05-2019/theme/a-balanced-and-progressive-trade-policy-to-harness-globalisation/file/multilateral-investment-court-(mic).

Concluding remarks

In the words of Commissioner Malmström: "The EU has been a key player in the reform process of ISDS in many ways: through participation in international fora (the UNCITRAL forum), through changes in the way new trade agreements are made with third countries, by judicial review of ISDS clauses in intra-EU Investor Agreements and a review of the new EU trade policy of creating a judicial forum for the resolution of investment disputes in new trade agreements. The European Commission stands ready to engage even more actively in the reform of the global arbitration framework and promote the new multilateral approach, following the recent CJEU Opinion 1/17".[41] Indeed opinion 1/17 clarified the legal boundaries of ISDS mechanisms under EU law and gave a new impetus to the ICS promoted by the EU in international trade and investment agreements. By endorsing ICS, the ECJ also gave a vote of confidence for the continuation of negotiations for the creation of a Multilateral Investment Court in the context of UNICTRAL. The set-up of the MIC could be a significant step forward in creating a stable international investment environment, which would be globally beneficial. Fostering investment stability through a fair, independent, dispute settlement system, that does not jeopardize the ability of the EU and its Member States to continue pursuing public policy objectives, is the key challenge that the EU has already identified and actively started addressing.

---

41 Speech by European Commissioner for Trade Cecilia Malmström, op. cit.

Agis Chr. Georgiades[*]

5 Energy dispute settlement after *Achmea* and opinion 1/17

Abstract: This chapter discusses the judgment of the Court of Justice of the European Union (CJEU) dated 06/03/18[1] in *Slowakische Republik v. Achmea BV* ("*Achmea*"), and opinion 1/17 dated 30/04/19.[2] The discussion focuses on the impact of these decisions to the mechanisms that have been established and used in the last few decades for the resolution of investment disputes, with special reference to disputes in the energy sector. An attempt is made to assess these decisions critically and to address the concerns they have raised in respect of the future of energy dispute settlement within the European Union (EU). It is shown that the prevailing interpretation of *Achmea* has caused much uncertainty at a time when the EU and its Member States do not have an effective alternative in place. It is probable that the CJEU will be called to revisit the matter. However, it is unlikely for the CJEU to take a different view from the prevailing interpretation of *Achmea*. This is because the stand taken by CJEU and other EU Institutions makes it clear that they do not intend to allow the resolution of intra-EU investment disputes to revert to the pre-*Achmea* era. Inevitably, the uncertainty caused in the resolution of energy and other investment disputes within the EU shall have an adverse impact on commerce and investment. It is doubtful whether the multilateral investment court sought to be introduced by EU institutions as a response to *Achmea*, will give a direct and adequate solution to these issues.

Keywords: *Achmea*, investment, BIT, arbitration, ECT, ISDS, multilateral

Introduction

The continuous influence of international arbitration to the contemporary economic environment is beyond doubt. Developing as the standard method for resolving international commercial and investment disputes, arbitration has

---

[*] Agis Chr. Georgiades, LLB, LLM, DipIM-ADR, DIN, FCIArb, CIArb Accredited Mediator, Barrister Partner, CGA Cyprus.
[1] European Court of Justice (Grand Chamber), judgment of 6 March 2018, case C-284/16, ECLI:EU:C:2018:158, *Achmea*.
[2] European Court of Justice, opinion 1/17 of 30 April 2019, ECLI:EU:C:2019:341, *CETA*; see also opinion of AG Bot delivered on 29 January 2019, ECLI:EU:C:2019:72, *CETA*.

arguably established an autonomous, international legal order.[3] This argument is supported by a clear trend to subject international dispute settlement procedures to instruments that do not form part of national legal orders. Examples of such instruments are the IBA Rules on the taking of evidence[4] and the more recently introduced Prague Rules.[5] In the same context, arbitration tribunals were gradually given the power to decide, at least at first instance, issues of arbitrability as well as issues regarding their own jurisdiction. These principles now form part of most developed legal systems. The decision in *Achmea* appears to come in direct conflict with this trend. This issue is examined below, with particular reference to intra-EU energy disputes.

1 Dispute settlement in bilateral and multi-lateral investment treaties

Aiming to encourage inbound investment, most states have entered bilateral or multilateral treaties with other states. Bilateral investment treaties for the mutual protection of investment are known as BITs. The first BIT was entered by Germany and Pakistan approximately sixty (60) years ago. Today, there are more than two thousand six hundred (2.600) BITs worldwide.[6] These typically include provisions for the protection of an investor of one state-party when investing in another state-party. Most BITs also include mechanisms for the resolution of disputes between an investor and a state-party, known as Investor-State Dispute Settlement ("ISDS") clauses. Usually, ISDS clauses state that disputes are to be resolved by institutional international arbitration. Thus, whereas international commercial arbitrations ordinarily derive from contractual arbitration clauses, investment arbitrations usually emanate from ISDS provisions in BITs or multi-lateral treaties. In relation to investment disputes, the Washington Convention ("ICSID Convention")[7] is of cardinal importance.

---

3   See generally: Emmanuel Gaillard, "Legal theory of International Arbitration" (Leiden/Boston, MA: Martinus Nijhoff Publishers, 2010).
4   Available at: https://www.ibanet.org/Document/Default.aspx?DocumentUid=68336C49-4106-46BF-A1C6-A8F0880444DC.
5   "Rules on the Efficient Conduct of Proceedings in International Arbitration (Prague Rules)", (2018), https://praguerules.com/prague_rules/.
6   See UNCTAD, "Investment Policy Hub", https://investmentpolicy.unctad.org/ (accessed 20 September 2019).
7   "ICSID Convention, Regulation and Rules", *International Centre for Settlement of Investment Disputes ICSID*, 15 April 2006, https://icsid.worldbank.org/en/Documents/resources/2006 %20CRR_English-final.pdf.

This is the treaty that established the International Centre for Settlement of Investment Disputes. One of the characteristics of ICSID arbitration is that an ensuing arbitration award becomes automatically enforceable in each member state and has the same effect as a judgment issued by the courts of such state. This characteristic is of importance to the *Achmea* discussion, as is shown below.

2 Energy Charter Treaty

International investment disputes often relate to the energy sector. Hence the possibility of invoking the protection provisions of one or more BIT, or those of multilateral treaties, is an essential factor to be considered when examining the making of such an investment. Perhaps the most important treaty in this regard is the Energy Charter Treaty ("ECT").[8] ECT is a multilateral treaty which was introduced with the intention of promoting cooperation between member states in the energy sector. It is widely invoked in the resolution of energy disputes. Approximately 30 % of investment arbitrations worldwide in 2016 related to the ECT.[9] Like most BITs, ECT contains an ISDS clause providing for international arbitration, and interestingly, the EU (and all its Member States) has also signed and ratified the ECT.

3 The *Achmea* case

In investment disputes, a conflict is thought to exist between EU law and intra-EU BITs and their ISDS clauses. In particular, it has been argued for several years that the ISDS clauses violated EU law by allowing arbitration tribunals to decide on issues of interpretation and application of the EU Treaties, for which EU Member States have agreed to give exclusive competence to the CJEU. Intra-EU BITs numbered approximately 200 in 2018. *Achmea* was not a dispute in the energy sector. However, it touched base on this conflict and has given rise to significant controversy amongst academics and practitioners with interest in investment arbitration. The *Achmea* case relates to the BIT between Slovakia and The Netherlands. The *Achmea* company had invested in the private insurance services sector in Slovakia. Being a subsidiary of a Dutch group of insurance companies, *Achmea* had been incorporated for the purposes of

---

8   Available at: https://energycharter.org/.
9   See UNCTAD, "Fact Sheet on Intra-European Union Investor-State Arbitration Cases", Issue 3, December 2018, https://unctad.org/en/PublicationsLibrary/diaepcb2018d7_en.pdf.

the said investment. The investment was triggered by the liberalization of the health insurance market in Slovakia in 2004. Two years later, after the investment was made by the Dutch investors through *Achmea*, Slovakia proceeded to legislative amendments that significantly affected the liberal regime in the Slovakian market, amongst others, by prohibiting the distribution of profits.[10] On this ground, *Achmea* referred a dispute to international arbitration in 2008, under the relevant BIT and the UNCITRAL Arbitration Rules. The seat of the arbitration was set in Frankfurt, and the arbitration was administered by the Permanent Court of Arbitration at The Hague.[11]

3.1 The arguments of Slovakia and the European Commission

The *Achmea* arbitration was concluded in 2012, with an award in favour of the applicants. The tribunal found that the relevant BIT had been violated by Slovakia and awarded approximately 22 million Euros as compensation to *Achmea*.[12] As a response, Slovakia applied to the Courts of Frankfurt with a request to set aside the award, arguing that the arbitration provision in the BIT violated Articles 18, 267 & 344 of the Treaty on the Functioning of the European Union ("TFEU").[13] This was not a new argument. It was a position repeatedly put by the European Commission for many years. The Commission had repeatedly asked to intervene in intra-EU investment arbitrations in order to pursue this position. However, arbitration tribunals systematically rejected such requests.[14] In *Achmea*, the basic position of Slovakia was that the ensuing legal questions related to fundamental provision of EU law, particularly in relation to the free movement of capital, which were superior to national law and the relevant BIT. Since the arbitration tribunal could not refer a matter for a preliminary ruling by the CJEU for interpreting these provisions, granting jurisdiction to the tribunal to interpret the provisions amounted to a violation of the prohibition on EU Member States from settling disputes relating to the interpretation or application of the Treaties in ways that are different to those

---

10 European Court of Justice (Grand Chamber), judgment of 6 March 2018, case C-284/16, ECLI:EU:C:2018:158, *Achmea*, para. 6–8.
11 Ibid., para. 9–10.
12 Ibid., para. 12.
13 Ibid., para. 11, 14.
14 Clément Fouchard and Marc Krestin, "The Judgment of the CJEY in Slovak Republic v. Achmea – A Loud Clap of Thunder on the Intra-EU BIT Sky!", *Kluwer Arbitration Blog*, 7 March 2018, http://arbitrationblog.kluwerarbitration.com/2018/03/07/the-judgment-of-the-cjeu-in-slovak-republic-v-achmea/.

provided for in the Treaties. For the same reasons, Slovakia argued that the BIT and the arbitration provision within it were contrary to public order and therefore void, and that giving effect to the arbitration provision would constitute unfair treatment in favour of the Dutch investor and to the detriment of investors from other Member States. The case reached the Federal Court of Germany, which referred questions for a preliminary ruling by the CJEU.[15] In the proceedings before the CJEU, several Member States intervened. States that are usually the respondents in investment arbitrations supported the positions of Slovakia. On the other hand, Member States from which the applicant-investors usually originate supported the contrary positions. This is indicative of the social and economic aspect of the matter.[16]

3.2 CJEU on *Achmea*

Advocate General Wathelet issued an opinion on 19/09/17.[17] The opinion is well-substantiated and detailed. With strong rhetoric, the Advocate General rejected the positions of Slovakia, but, nevertheless, the CJEU took a different view. The Judgment of the CJEU was issued on 06/03/18.[18] In summary, the Court considered that Member States undertook the obligation not to submit disputes concerning the interpretation of the application of the EU Treaties to any method of settlement other than those provided for in the EU Treaties.[19] As an arbitration tribunal is not a court and therefore cannot refer a question of EU law to the CJEU for a preliminary ruling,[20] the arbitration provisions

---

15 European Court of Justice (Grand Chamber), judgment of 6 March 2018, case C-284/16, ECLI:EU:C:2018:158, *Achmea*.
16 Deyan Dragiev, "A Procedural Perspective of Achmea: What Does Achmea Imply in Practice", *Kluwer Arbitration Blog*, 10 June 2018, http://arbitrationblog.kluwerarbitration.com/2018/06/10/procedural-perspective-achmea-achmea-imply-practice/; Vivek Kapoor, "Slovak Republic v. Achmea: When Politics Came Out to Play", *Kluwer Arbitration Blog*, 1 July 2018, http://arbitrationblog.kluwerarbitration.com/2018/07/01/slovak-republic-v-achmea-politics-came-play/.
17 Opinion of AG Wathelet delivered on of 19 September 2017, case C-284/16, ECLI:EU:C:2017:699, *Achmea*.
18 European Court of Justice (Grand Chamber), judgment of 6 March 2018, case C-284/16, ECLI:EU:C:2018:158, *Achmea*.
19 "Treaty on the Functioning of the European Union", OJ C 115, 9 May 2008, p. 47, Article 344.
20 Ibid., Article 267.

in intra-EU BITs and other treaties are unenforceable. The Court found that the arbitration tribunal that had been established based on the provisions of the relevant BIT could not be considered a 'court' for the purposes of Article 267 of the TFEU. This was because it lacked enough juridical connection to the Member States that had entered the BIT. Therefore, it lacked the power to directly refer questions for a preliminary ruling to the CJEU.[21]

The CJEU noted that such referral would only be possible by the court of the seat of the arbitration, in the event that an application to set aside the arbitration award was filed. However, that Member State's court's powers to review the arbitration award would be limited to those prescribed by the law of that Member State. In *Achmea*, German law provided very limited powers to German courts to review the arbitration award, mainly in relation to the validity of the arbitration provision of the BIT and rules of public order. Hence the possibility to refer a question to the CJEU for a preliminary ruling was substantially limited to such and related matters.[22] The conclusion of the CJEU was that the interpretation of the particular provisions of the TFEU prohibited the inclusion in international treaties between Member States – such as the relevant BIT – of mechanisms by which an investor from one Member State investing in another Member State could refer to arbitration a dispute relating to the investment. It is important that the CJEU reached this conclusion without considering whether the arbitration tribunal in the case in question had interpreted or applied the EU Treaties. For the CJEU, it was enough that a tribunal in such a place could *potentially* be called to interpret or apply the Treaties.[23] The CJEU acknowledged that such a possibility is also present in international commercial arbitrations. But it considered the case for investment arbitration to be different. In particular, it found the difference in that commercial arbitrations arise from the parties' free will, whereas investment arbitration arises from acts of the Member States. What makes investment arbitration different according to the CJEU is that the possibility to refer a dispute to arbitration arises from a positive act of the Member States, that is the entering and ratification of a BIT

---

21 European Court of Justice (Grand Chamber), judgment of 6 March 2018, case C-284/16, ECLI:EU:C:2018:158, *Achmea*, para. 46–49.
22 Ibid., para. 52–56.
23 Ibid., para. 60; see also: Volodymyr Ponomarov, "CJEU Does not Buy Wathelet's Opinion in Achmea – What is Left Unanswered?", *Kluwer Arbitration Blog*, 14 April 2018, http://arbitrationblog.kluwerarbitration.com/2018/04/14/cjeu-not-buy-wathelets-opinion-achmea-left-unanswered/.

with an ISDS clause providing for arbitration.[24] On these grounds, the CJEU considered that the putting in place of such a treaty between the Member States potentially equals to the settling of disputes relating to the interpretation or application of the EU Treaties in a way that is different to that prescribed in the Treaties themselves. Hence, the CJEU held that the BIT in question was potentially relevant to the interpretation and application of the Treaties, and therefore, the ISDS clause providing for arbitration was incompatible to Articles 267 & 344 TFEU. The CJEU did not consider the argument for unfair treatment. The CJEU diagnosed a risk in allowing arbitration tribunals to interpret and apply the EU Treaties since the findings of such tribunals cannot be reviewed by the CJEU through the procedure of the preliminary ruling. But this diagnosis occurred in a procedure for a preliminary ruling and invariably lead to the setting aside of the award by the German court.[25] It is, therefore, reasonable for one to question the seriousness of the risk, given that the risk was identified by the CJEU in the *Achmea* Judgment, which resulted in the annulment of the *Achmea* award. One could argue that the ultimate result of the *Achmea* judgment refutes the risk identified in the same judgment.

4 Opinion 1/17 of the CJEU on CETA

The Investment Court established by the EU-Canada Comprehensive Economic and Trade Agreement (CETA) gives a clear indication of the intentions of European institutions.[26] It is evident that the EU intends to replace the ISDS mechanisms of BITs with tribunals within the control of EU institutions. To this end, the EU obtained the green light by opinion 1/17 of the CJEU, which followed the view of Advocate General Bot and found the tribunal provision in CETA is compatible with EU law.[27]

---

24 Ibid., para. 55.
25 Tom Jones, "Germany's Top Court Shows Obedience to Achmea", *Global Arbitration Review* (2018).
26 See European Court of Justice, opinion 1/17 of 30 April 2019, ECLI:EU:C:2019:341, *CETA*; see for an overview Press release No. 52/19 of 30 April 2019, EU:C:2019:341, *The mechanism for the resolution of disputes between investors and States provided for by the free trade agreement between the EU and Canada (CETA) is compatible with EU law*, https://curia.europa.eu/jcms/upload/docs/application/pdf/2019-04/cp190052en.pdf.
27 European Court of justice, opinion 1/17 of 30 April 2019, ECLI:EU:C:2019:341, *CETA*; opinion of AG Bot delivered on 29 January 2019, ECLI:EU:C:2019:72, *CETA*.

5 The prevailing interpretation of *Achmea*

Just two weeks after the *Achmea* judgment was issued, on 20/03/18, the European Council authorised[28] the Commission to negotiate the preparation of a treaty for a multilateral investment court, for the settlement of intra-EU investment disputes. This attempt received severe criticism. Amongst others, a concern was raised that, since the judges of this court will be appointed by the governments of Member States, the usual respondents will have the exclusive saying in making the appointments. On the other hand, investors, who are the usual claimants in investment disputes, will not have a saying in the appointments at all.[29] In a communication by the European Commission to the European Parliament dated 19/07/18, it is expressly stated that the judgment in *Achmea* is relevant to the ECT. In particular, the communications sated that the arbitration provision in ECT could not be applied to intra-EU investment disputes.[30] Furthermore, in a common declaration of the representatives of Member Stated dated 15/01/19,[31] Member States declared that they recognize the results of *Achmea*, amongst others, that *Achmea* rendered inapplicable the arbitration provisions in intra-EU BITs and the ECT. They also undertook to revoke the relevant BITs.[32]

In *Achmea*, it was considered that the relevant BIT provided for the application of the law of Slovakia to the merits of the dispute. Since the law of Slovakia incorporated that of the EU, the CJEU considered that there was a potential for

---

28 See http://data.consilium.europa.eu/doc/document/ST-12981-2017-ADD-1-DCL-1/en/pdf.
29 Marike R. P. Paulsson, "Revisiting the Idea of ISDS Within the EU and an Arbitration Court: The Effect on Party Autonomy as the Main Pillar of Arbitration and the Enforceability of Arbitral Awards", *Kluwer Arbitration Blog*, 21 May 2018, http://arbitrationblog.kluwerarbitration.com/2018/05/21/revisiting-idea-isds-within-eu-arbitration-court-effect-party-autonomy-main-pillar-arbitration-enforceability-arbitral-awards/.
30 European Commission, Protection of intra-EU investment, COM(2018) 547 final of 19 July 2018, pp. 3–4.
31 Declaration of the Member States of 15 January 2019 on the legal consequences of the Achmea judgment and on investment protection, 17 January 2019, https://ec.europa.eu/info/sites/info/files/business_economy_euro/banking_and_finance/documents/190117-bilateral-investment-treaties_en.pdf.
32 Deyan Dragiev, "2019 in Review: Investment Arbitration in Europe", *Kluwer Arbitration Blog*, 9 January 2020, http://arbitrationblog.kluwerarbitration.com/2020/01/09/2019-in-review-investment-arbitration-in-europe/?print=print.

the arbitration tribunal to he called to interpret and apply the Treaties. From this consideration, emanated a counter-argument that intra-EU BITs that do not provide for the direct or indirect application of EU law are unaffected by *Achmea*. Thus, the prevailing interpretation of *Achmea* is somewhat arbitrary. Firstly, because Article 26(3)(a) ECT provides expressly that each contracting party (i.e. including all EU Member States and the EU itself) gave its unconditional consent to the submission of a dispute to international arbitration. And secondly, because Article 26(6) ECT, unlike the relevant BIT in *Achmea*, provides that issues in dispute are to be decided in accordance with the ECT and principles of international law and not under the law of a state or the EU. These arguments are supported by the findings in an ICSID case that followed *Achmea*, *Up and C.D. Holding Internationale v. Hungary*,[33] decided under the France-Hungary BIT. In this arbitration, Hungary raised the issue of jurisdiction relying on *Achmea*. The arbitration tribunal rejected the objection, considering as its jurisdictional base the Washington Convention. A related argument was examined in the ICSID arbitration *Vattenfall and other v. Germany*, which resulted in a final ward on 31/08/18.[34] This case concerned an intra-EU investment dispute under the ECT. The arbitration tribunal considered there was a risk that a final award would not be enforced within the EU. But nonetheless, it considered that this risk could not justify a refusal by the tribunal to embark upon and conclude the procedure that was assigned to it under the ECT. The arbitrators decided that the issue of the validity of the arbitration provision in the ECT should be decided under rules of public international law and rules relating to international contracts, and not under rules of EU law. They, therefore, concluded that they had jurisdiction to hear the claim, refusing to apply *Achmea* on the facts of the case before them. Similar decisions were reached in subsequent awards.[35] Irrespective of the above, it is difficult to trace a tendency to overturn *Achmea*. It would be rather difficult for a court of a Member States

---

33 ICSID, award of 9 October 2018, case No. ARB/13/35, *UP (formerly Le Chèque Déjeuner) and C.D Holding Internationale v. Hungary*, https://www.italaw.com/sites/default/files/case-documents/italaw10075.pdf.
34 ICSID, decision of 31 August 2018, case No. ARB/12/12, *Vattenfall AB and others v. Federal Republic of Germany*, https://www.italaw.com/sites/default/files/case-documents/italaw9916.pdf.
35 Markus Beham and Désirée Prantl, "Intra-EU Investment Reform: What Options for the Energy Charter Treaty?", *Kluwer Arbitration Blog*, 7 January 2020, http://arbitrationblog.kluwerarbitration.com/2020/01/07/intra-eu-investment-reform-what-options-for-the-energy-charter-treaty/.

to consider a provision for arbitration in an intra-EU BIT to be valid. Hence the possibility of enforcing an ensuing award within the EU is very remote. It will also be interesting to see how ICSID awards relating to intra-EU investment disputes will be seen. Where the seat of such dispute is outside the EU, it is unlikely for the supervising court to set aside an ensuing award on the grounds of *Achmea*. The award will then have to be enforced, despite being an intra-EU investment award. Refusal to allow its enforcement would contravene the ICSID Convention.

6 Criticism against *Achmea*

A significant criticism against the judgment in *Achmea* is that it disregarded certain pivotal mechanisms of international arbitration. It did not consider the principle of arbitrability, according to which arbitration tribunals may, using substantial available case-law, examine whether the issues in dispute can be settled by arbitration. Also, *Achmea* did not take into account the widely recognized principle of 'Kompetenz-Kompetenz', which prescribes that tribunals may decide, at least at first instance, on their own jurisdiction. These mechanisms can be used by parties and tribunal in investment arbitrations in order to determine, early in the settlement process, whether the dispute in question raises matters that are beyond the jurisdiction of the tribunal or are not arbitrable for reasons of public policy, by application of EU law or otherwise. In other words, arbitration has developed autonomous mechanisms by which it can address the concerned raised in the *Achmea* judgment. Had the CJEU considered these matters, the judgment in *Achmea* may have been decided differently. Furthermore, the CJEU in *Achmea* did not take into account the possibility to refer issues of law to a competent state court before or during arbitration proceedings. In England and Wales, this is possible by reference to s.32 of the Arbitration Act 1996,[36] which establishes a mechanism by which a court may determine any question as to the substantive jurisdiction of the tribunal.

One could argue that the concern in *Achmea* (i.e. that arbitration tribunals should not be allowed to interpret the EU Treaties) is wider in scope. On the other hand, other jurisdictions retain provisions allowing much wider court intervention before or pending arbitration proceedings. The example of Cyprus is perhaps the most illustrative on this point. Cyprus has a dual *lex arbitri*. International commercial arbitration is governed by the Cyprus International

---

36 Arbitration Act 1996, http://www.legislation.gov.uk/ukpga/1996/23/contents.

Commercial Arbitration Law of 1987.[37] This law is based on the UNCITRAL Model Law on International Commercial Arbitration.[38] But it does not appear to apply to investment arbitrations seated in Cyprus, because such arbitrations are not 'commercial'. It, therefore, seems that an investment arbitration seated in Cyprus would be governed by the rather archaic Cyprus Arbitration Law,[39] a statute which was enacted in 1944, has never been amended, and is thought to apply mainly to domestic arbitration proceedings in Cyprus. This archaic law retains the case-stated procedure, according to which tribunals may – and are obliged if all parties agree – to refer any legal matter to the court for a decision. Therefore, if the *Achmea* arbitration was seated in Cyprus, the tribunal would most probably have to refer all legal matters to a local court. And that local court could then refer matters relating to the interpretation or application of the EU Treaties for a preliminary ruling by the CJEU. It follows that *Achmea* should have been confined to its own facts and circumstances. The main reasoning of the CJEU (i.e. that an arbitration tribunal cannot refer a question for a preliminary ruling) is not as wide as the prevailing interpretation of *Achmea* seems to suggest. Another point to consider is that EU Member States can amend their arbitration laws so that they can all include provisions obligating arbitration tribunals to refer issues relating to the interpretation or application of the EU Treaties to state courts. In fact, this can even be done by primary or secondary EU legislation. That EU Member States are also members of the Convention on the Recognition and Enforcement of Foreign Arbitral Awards (New York Convention) is not an obstacle to such amendment as the said convention regulates the enforcement of awards and not setting aside of awards or court intervention before or during arbitration proceedings. Given the above, it is suggested that the CJEU may be called to revisit the issue of intra-EU investment arbitration provisions. However, given the apparently strong EU policy towards terminating these provisions, it is doubtful whether the CJEU confine *Achmea* to its own facts and circumstances.

---

37   Cypriot Law 101/1987, http://www.cylaw.org/nomoi/enop/non-ind/1987_1_101/full.html.
38   UNCITRAL Model Law on International Commercial Arbitration, United Nations, 1994, https://www.uncitral.org/pdf/english/texts/arbitration/ml-arb/06-54671_Ebook.pdf.
39   Cypriot Law, Arbitration Law, Chapter 4, http://www.cylaw.org/nomoi/enop/non-ind/0_4/full.html.

7 Adequacy of EU response

As mentioned above, January 2019, the EU Member States adopted declarations by which they undertook to terminate intra-EU BITs and to establish a single EU regulation and investment tribunal for intra-EU investment disputes. On 24/10/19, the European Commission announced that the EU Member States agreed on a plurilateral treaty for the termination of all intra-EU BITs. According to this agreement, all intra-EU BITs shall be terminated by the operation of the agreement. Sunset clauses contained in intra-EU BITs are made devoid of legal effect. Arbitration clauses contained in intra-EU BITs are nullified. However, concluded investment treaty arbitrations are not prejudiced by the agreement. Hence, the agreement shall not alter the results of disputes which were already completed, as the CJEU judgment did in the *Achmea* case. But for pending disputes, the agreement provides that they shall be resolved via so-called structured dialogue. The lack of clarity in these provisions raises substantial concerns as to how it will operate in practice.[40] As for the investment tribunal sought to be introduced at EU level, several significant questions remain unanswered. Amongst others, it is not clear if potential investors will be involved in the setting up the tribunal, its rules or composition. It is also unclear whether the tribunal will apply the principles of international law that have developed in investment arbitration cases, or if the relevant jurisprudence of several decades will be disregarded. Many practical and substantial problems are anticipated to arise in cases involving investors of dual nationality, one being non-EU, or investors that are joint ventures between EU and non-EU parties. Perhaps most importantly, it is not known how the decisions of such tribunal will be recognized and enforced outside the EU, if at all.

Conclusions

It is obvious that strong policy requires a different method of settling investment disputes in the EU. Irrespective of *Achmea*, investment arbitrations continue to be filed and pursued. In fact, 2018 saw an increase in such cases.[41] It is

---

40 Deyan Dragiev, "2019 in Review: Investment Arbitration in Europe", *Kluwer Arbitration Blog*, 9 January 2020.
41 See IBA Arbitration Subcommittee on Investment Treaty Arbitration, "Consistency, efficiency and transparency in investment treaty arbitration", October 2018, https://uncitral.un.org/sites/uncitral.un.org/files/investment_treaty_report_2018_full.pdf.

beyond doubt that *Achmea* came at a time when there is no alternative mechanism in place. This has caused and shall continue to cause uncertainty and have an adverse impact on transactions and intra-EU investment and synergies in the energy sector.

Louise Bouvery[*]

6 The use of technically based alternative dispute resolution mechanisms in oil & gas contracts

Abstract: The oil & gas sector is not estranged to private and alternative dispute resolution mechanisms (ADRMs), i.e. outside of national courts. This is due to, among others, the technicality of the disputes, the need for neutrality and efficiency. However, the ever-growing resort to arbitration tends to deprive it of its original characters of time and cost-efficient mechanism. Practice needs, therefore, to adapt and present new systems of dispute resolution. The aim of efficient cost-wise and time-wise ADRMs could be achieved by the implementation of three different mechanisms: Technical Advisory Committees; Dispute Review Boards; and Sole Expert Determination. These mechanisms are two-pronged as they act not only as Dispute Resolution Mechanisms (DRM) but as Dispute Prevention Mechanisms (DPM). This chapter will present and analyse these three proposed 'unconventional' ADRMs and discuss the benefits of implementing them in oil and gas contracts, as an alternative or complementary step to arbitration.

Keywords: Alternative Dispute Resolution Mechanisms, arbitration, Technical Advisory Committees, Dispute Review Boards, Sole Expert Determination

Introduction

The use of Alternative Dispute Resolution Mechanisms is not new in the oil & gas sector. These methods, including mediation, expert determination or arbitration, among others, appear safer to the industry's actors. As opposed to national courts, the third party involved in the dispute resolution process is an expert whose impartiality and neutrality is less questionable. Arbitration has been the preferred choice of dispute resolution method in many sectors. Most Bilateral Investment Treaties, regional treaties – such as the Energy Charter Treaty –, and investment and commercial contracts, provide for arbitration instead of resorting to national courts.[1] Arbitration presents many benefits

---

[*] PhD Candidate, University of Exeter (UK).
[1] Bernardo M. Cremades and David J. A. Cairns, "The Brave New World of Global Arbitration", *Journal of World Investment 3,* no. 2 (2002), 175–177.

to the parties. They can choose their arbitrators, the rules applicable to the procedure, the language, the seat, but, more so, they can ensure that the procedure will be impartial and neutral. This is especially true regarding oil & gas contracts as they involve State-owned natural resources and the State as a party to the contract. In addition, arbitration is internationally recognised and enforceable.[2] Finally, with arbitration being a private dispute resolution mechanism, parties are not subject to courts' calendars. Such appealing characteristics have created a surge in the use of arbitration as a dispute resolution mechanism. At its start, arbitration was promoted as a time and cost-efficient mechanism, freed from the constraints of national litigation. It is, however, recognised today as costly and time-consuming, to the point that it has sometimes been referred to as the new litigation.[3] This does not mean that arbitration must be abandoned, but rather that its use must be limited to complex disputes, where other mechanisms are helpless. This role of "filter" has to be played by Alternative Dispute Resolution Mechanisms. Some, like mediation or negotiations, are more well known while others, like, for instance, Technical Advisory Committees and Dispute Review Boards, are more obscure. Decreasing the number of disputes sent to arbitration can be obtained by solving the disputes at an earlier stage. In this context, technical-based ADRM or 'unconventional' ADRM may have a crucial role. Technical-based ADR encompasses Technical Advisory Committees, Dispute Review Boards and Sole Expert Determination. A crucial aspect of these ADRMs is that they are implemented prior to the existence of a dispute, or directly at its emergence. More importantly, the dispute resolution process for the three mechanisms relies on one or several experts. As opposed to other mechanisms, 'unconventional' ADRMs are efficient as they can settle both technical and legal disputes. Moreover, they act not only as dispute resolution tools but as dispute prevention methods as well. The three mechanisms will be presented, before assessing their effectiveness *in lieu* of arbitration as well as a previous step to arbitration. The question of implementing them in a multi-tiered dispute resolution clause will then be discussed.

---

2   New-York Convention on the Recognition and Enforcement of Foreign Arbitral Awards (1958).
3   Thomas J. Stipanowich, "Arbitration: The New Litigation", *University of Illinois Law Review 2010*, no. 1 (2010), 1.

1 The new methods: Technical Advisory Committees, Dispute Review Boards, Sole Expert Determination

1.1 General characters of the mechanisms

1.1.1 Technical Advisory Committees

The Technical Advisory Committee (hereafter TAC) is a technical assembly in which the parties to the Exploration and Production programme meet and discuss findings, future plans for the performance of the contract. Often seen as a technical tool only, the TAC bears characteristics that make it an Alternative Dispute Resolution Mechanism. TACs first emerged as Joint Management Committees. Joint Management Committees arose from the shift from concessions contracts to Production Sharing Agreements (PSA) in Indonesia. In order to balance its loss of ownership rights over the petroleum resources under the PSAs, Indonesia implemented "management" clauses in the agreements.[4] These clauses provided for the submission of work programmes and budget to the Indonesian National Oil Company, Pertamina, by the International Oil Company.[5] Once the commercial production phase was reached, the International Oil Company and Pertamina had a monthly "operating" meeting.[6] TACs are common in oil & gas investment contracts, albeit bearing different denominations. In the Azerbaijan PSA, the Committee exists under the name of Steering Committee.[7] The 1993's and 2000's PSAs of Nigeria provide for a Management Committee.[8] Finally, the Greek Model Lease

---

4 Zhiguo Gao, "International Petroleum Contracts: Current Trends and New Directions" (Graham and Trotman, 1994), pp. 80–81.
5 Robert Fabrikant, "Production Sharing Contracts in the Indonesian Petroleum Industry", *Harvard International Law Journal* 16, no. 2 (1975) 303, 312.
6 Fabrikant, "Production Sharing Contracts", cit., p. 315.
7 Agreement on the Joint Development and Production sharing for the Azeri and Chirag fields and the deep water portion of the Gunashli field in the Azerbaijan sector of the Caspian Sea, concluded between the State Oil Company of the Azerbaijan Republic and ten International Oil Companies (Azerbaijan PSA), Article 5.2 (a). This contract has been provided by TOGYiN, The Oil and Gas Year, https://www.togyin.com.
8 Mohammed Babangida Umar, "Legal Issues in the Management of Nigeria's Production Sharing Contracts from a Study of the Nigerian National Petroleum Corporation's (National Petroleum Management Services') Perspective", *Oil, Gas & Energy Law Journal (OGEL)* 3, no. 1 (2005), 34.

Agreement for Exploration and Hydrocarbons in Greece establishes a TAC.[9] Despite the variety in the appellations, the term TAC will be retained here. On the issue of functions of the TAC, it needs to be said that TACs are composed of an equal number of representatives for both parties to the contract, in order to favour balanced discussions on the matters submitted to the appreciation of the TAC. The Oil Company is expected to present the work programme carried out since the last session and the work programme and budget for the next six months. Programmes and budget are discussed by the parties, alongside the results of any discoveries and plans for appraisal and development.[10] There are usually no limitations on the people to be appointed by the parties as serving members of the TAC. As such, the members of the TAC can be technicians, scientists or lawyers. However, as the discussions will focus mainly on technical aspects of the exploration and production, one has to bear in mind that the members of the TAC must be able to understand the full range of the issues at hand. In order to guarantee that the largest positions and opinions are shared during the meetings, a certain quorum has to be respected. In the Greek Lease Model Agreement, for instance, the quorum has been set at five members.[11]

Furthermore, to favour global discussions and approval, the decisions have to be taken unanimously.[12] The TAC meets regularly to assure that all matters are considered and settled in due time. The regularity of these meetings is provided in the contract. This does not, however, prevent the parties to decide to meet more regularly. Furthermore, as to ensure the settlement of matters or issues in a timely manner, each party to the contract can convene an extraordinary meeting of the TAC, for emergency and extraordinary situations, save a prior notification.[13] On the issue of aims, duties and powers of the TAC, it needs to be said that a TAC has as the main role to engage on different subjects pertaining to the good performance of exploration and production of hydrocarbons, in line with the contract provisions. The functions of the TAC are both an approval/review role and a monitoring role. As such, the TAC will review the Annual Work Programme and Budget; any Appraisal Programme; and any Development and Production Programme. This discussion/negotiation role of

---

9   Model Lease Agreement, Article 4, Athens, December 2017, https://www.greekhydrocarbons.gr/pdfs/BlockDocuments/Model_Lease_Agreement.pdf.
10 Michael A.G. Bunter, "The Promotion and Licensing of Petroleum Prospective Acreage", *Kluwer Law International* (2002), 270 et seq.
11 Model Lease Agreement, Article 4.7.
12 Model Lease Agreement, Article 4.9.
13 Model Lease Agreement, Article 4.6; Azerbaijan PSA, Article 5.2 (i).

the TAC demonstrates its crucial role in the prevention and solving of any possible issues that may arise from the performance of exploration and production of hydrocarbons.

1.1.2 DRBs

DRBs are Alternative ADRMs mainly provided in construction contracts, but are not yet common in the oil & gas industry despite their recognised effectiveness.[14] DRBs are the consequence of the lengthy and costly litigations in the construction industry, and the realisation that the "traditional" ADR (i.e. arbitration, mediation and mini-trials among others) were only available *ex post* the completion of the project.[15] DRBs are constituted of three independent members chosen by the parties for their knowledge and their professionalism. Boards are appointed at the conclusion of the construction contract and regularly visit the construction site, or at critical stages of the construction.[16] These periodic visits allow the members to become familiar with the operations and the parties involved,[17] and put them at the forefront to hear unresolved disputes, as soon as they arise.[18] These characteristics of the DRBs concur in establishing legitimacy and a relationship between the Board and the parties. Regarding the procedure, hearings and recommendations, it needs to be mentioned that procedure before a DRB is informal and non-adversarial.[19] Once the parties have come to the conclusion that a settlement is unlikely, they can submit their claim to the Board which will arrange hearings.[20] In order to guarantee the non-adversarial character of DRBs, all the parties to the dispute have to be

---

14 A. Timothy Martin, "Dispute resolution in the international energy sector: an overview", *Journal of World Energy Law and Business* 4, no. 4 (2011), 339.
15 Kathleen M. J. Harmon, "Effectiveness of dispute review boards", *Journal of construction engineering and management* 129, no. 6 (2003), 674.
16 John D. Coffee, "Dispute Review Boards in Washington State", *Arbitration Journal* 43, no. 4 (1988), 60.
17 Kathleen M. J. Harmon, "Construction conflicts and dispute review boards: Attitudes and opinions of construction industry members", *Dispute Resolution Journal* 58, no. 4 (2003), 68.
18 Roxene M. Thompson, Michael C. Vorster and James P. Groton, "Innovations to manage disputes: DRB and NEC", *Journal of Management in Engineering* 16, no. 5 (2000), 52.
19 Kathleen M. J. Harmon, "Resolution of construction disputes: A review of current methodologies", *Leadership and Management in Engineering* 3, no. 4 (2003), 196–197.
20 Harmon, "Resolution of Construction Disputes", 197.

involved; otherwise the process is deemed to fail.[21] The Board will then, generally, issue non-binding written recommendations which can be rejected by the parties but are admissible in further proceedings such as arbitration.[22] The recommendation is reasoned like an arbitration award or a judgment, which is highly appreciated by professionals.[23] The non-binding character of the DRB does not affect its efficacy; on the contrary, DRBs have a very high success rate of more than 90 %.[24]

1.1.3 Sole Expert Determination

SED differs from the previously exposed ADR methods (i.e. TAC and DRB) as this mechanism is adjudicative. The parties decide in the contract, or when the dispute arises, that they will refer their dispute to the opinion of an independent expert in whose hands the authority power lies.[25] The use of Sole Experts is common in the oil & gas sector for issues mostly related to technical matters, such as commercial reserves, the Development Plan, unitisation issues for the determination or re-determination of interest in oil & gas field, or the application of price review provisions.[26] SED is often favoured because it involves an expert in highly technical matters, giving conviction to the parties that the issue will be handled with great professionalism. Moreover, the appointment of

---

21 Daniel D. McMillan, "An Owner's Guide to Avoiding the Pitfalls of Disputes Review Boards on Transportation Related Projects", *Transportation Law Journal* 27, no. 2 (Spring 2000), 199.
22 Daniel D. McMillan and Robert A. Rubin, "Dispute Review Boards: Key Issues, Recent Case Law, and Standard Agreements", *Construction Lawyer* 25, no. 2 (Spring, 2005), 14.
23 Macmillan and Rubin, "Dispute Review Boards", cit., 15; Harmon, "Construction conflicts and dispute review boards", cit., 74.
24 Colleen A. Libbey, "Working Together while Waltzing in a Mine Field: Successful Government Construction Contract Dispute Resolution with Partnering and Dispute Review Boards", *Ohio State Journal on Dispute Resolution* 15, no. 3 (2000), 844; Christopher T. Horner, "Should Dispute Review Board Recommendations Be Considered in Subsequent Proceedings", *Construction Law* 32, no. 3 (2012), 1; Donald Charrett, "The Commercial Value of Dispute Boards under FIDIC Contracts", *Turkish Commercial Law Review* 1, no. 3 (2015), 217–218.
25 Robert Hunt, "Cost effective resolution of construction disputes: wishful thinking or emerging reality? (Part 2)", *Asian Dispute Review* 4, no. 1 (2002), 95.
26 Michael A. G. Bunter, "The Promotion and Licensing", 284 et seq.; Henry Brown and Arthur Marriott, "ADR: Principles and Practice", (London: Sweet & Maxwell, 2011), 142, para. 7–037.

a Sole Expert is achieved conjointly by the parties, increasing impartiality, and therefore confidence, in the procedure. They are opposed to party-appointed experts, regularly used before national courts and arbitral tribunal for cases involving a high level of science or engineering expertise, in which judges or arbitrators may not have the substantive and necessary knowledge. Party-appointed experts are often considered partisans ("hired-guns") as they are paid by the parties themselves, and their opinion may appear not reliable. Parties will also tend to appoint the best-spoken experts, the most convincing rather than a competent and recognised expert who will not appear as clear and convincing during cross-examination.

Regarding the nomination of the Sole Expert, the choice of the Sole Expert relies on the parties. They have to both agree on the person to act as the Sole Expert, although some contracts may provide specific guidelines. For instance, the Greek Model Lease Agreement gives some requirements regarding the person that will be the Expert. The Sole Expert under this Lease Agreement must indeed be a member of either the Energy Institute of London, the American Petroleum Institute or the French Institute of Petroleum.[27] Moreover, possible conflicts of interests are tackled in the Lease Agreement, by allowing the Lessor (the Hellenic Republic), to appoint another institute in this instance. The institute must originate from another European Union's member state that produces hydrocarbons.[28] The nomination of the Sole Expert relies on the agreement of both parties on the name. Hence SED clauses must account for a conflict between the parties on this choice. The Lease Agreement for the exploration and exploitation of hydrocarbons in the Southwest of Crete Block (among others) gives an alternative to the choice of the Sole Expert by the parties in case of a disagreement.[29]

In this instance, the nomination of the Sole Expert will be left to the President of one of the aforementioned institutes (Energy Institute of London, the American Petroleum Institute or the French Institute of Petroleum), until an Expert has been chosen. This externalisation of the nomination in the case of a conflict, protects the impartiality and neutrality of the procedure. It also allows to unlock the situation and to progress with the resolution of the first

---

27 Model Lease Agreement, Definitions.
28 Model Lease Agreement, Definitions.
29 Lease Agreement for the Exploration, Discovery and Production of Hydrocarbons, Block South West of Crete, *Official Government Gazzette*, 10 October 2019, Vol. 153, Article 23.2. (b), https://www.greekhydrocarbons.gr/pdfs/BlockDocuments/N.4628.pdf.

dispute. Regarding the credentials of the Sole Expert, contracts can be silent or very detailed. However, as for arbitrators, the qualities of the Sole Expert shall not be excessive, as it may render the nomination of an Expert fitting all the criteria very difficult if not impossible. The parties should be aware of the nature of the dispute they are referring to SED. If it relates to an unitisation, choosing an expert who is experienced and recognised in this domain would be preferable. The parties may also, if the contract allows it, choose an Expert who will have a law background in addition to its technical skills and experience. This can be of importance when both technical and legal matters are intertwined. The procedure of SED will depend on the content of the contract. Some may leave it all to the appreciation of the Sole Expert, with limits only as to the length of the procedure, while others may provide strict procedural rules. The Sole Expert will be given an outline of the dispute by the parties. This outline will correspond to the scope of the mission of the Sole Expert. It is worth noting that the Sole Expert cannot deviate from the scope he/she has been given by the parties. As such, if the dispute submitted to him/her relates to a Development Plan, the Sole Expert is not empowered to hear the parties on any other subject, might be relevant to the dispute or directly linked to it. From this limited scope, stems the risk of having the same dispute to be solved before two different forums: The aspect of the dispute comprised in the scope before the Sole Expert, and the remaining aspects before an arbitral tribunal. Parties should, therefore, be careful to include all sides of the dispute when giving a mission to the Sole Expert if they want to avoid an incomplete solving of their conflict.

The Sole Expert relies on the material presented by the parties he or she deems necessary and relevant and on its expertise.[30] He/she is also limited by the parties' submissions as he/she cannot compel them to provide documents.[31] The decision of the Sole Expert, therefore, depends on the willingness of the parties to provide the relevant documentation and evidence for the dispute to be solved. A party withholding evidence may endanger the dispute resolution process and lead the Sole Expert to issue a decision while not fully informed. On this note, the Greek Model Lease Agreement imposes on the parties an obligation to act in good faith and cooperate during the SED process.[32] Finally,

---

30 Robert Hunt, "Cost-effective resolution of construction disputes", cit., 96; Edwin H. W. Chan, Charles K. L. Chan and Martyn J. Hills, "Construction industry adjudication: A comparative study of international practice", *Journal of International Arbitration* 22, no. 5 (2005), 366.
31 Chan et al, "Construction industry adjudication", cit., 366.
32 Model Lease Agreement, Article 23.1 (i).

the decision of a Sole Expert is generally binding and definitive, which makes the process efficient as it applies immediately to the parties. Furthermore, contracts often provide for a timeline within which the Sole Expert has to issue its decision. This timeframe is kept short, as the aim of SED is to solve the dispute quickly. However, as opposed to a court judgment or an arbitration award, the determination only has a contractual binding power between the parties and cannot be directly and internationally enforced before national courts. Nevertheless, parties can seek the enforcement of the Sole Expert's decision through national courts, as they would do for any other contractual obligation. Despite the lack of internationally recognised binding force, SED is a valuable Alternative Dispute Resolution Mechanism that ought to be implemented largely in oil & gas investment contracts. SED does not bring the contract, i.e. the rights and obligations of the parties to a halt during the procedure. The exploration or exploitation operations can, therefore, continue to progress while the dispute at stake is being solved.[33] In addition, the procedure of SED is fast and cost-effective as parties normally share the costs of the Sole Expert. Although, and as opposed to TACs and DRBs, Sole Experts are not appointed from the inception of the contract, their nomination intervenes almost immediately after the dispute arises.

1.2 Dispute prevention

Traditional ADRMs, such as mediation, conciliation or arbitration, are only activated once the dispute has arisen. Their role is, therefore, solely one of dispute resolution. On the contrary, the three aforementioned ADRMs are also Dispute Prevention Mechanisms. This means that their role is two-pronged. They firstly act as intended, i.e. as dispute resolution tools to be used when a dispute arises, but, by their mere existence or their processes, also prevent disputes to be created. The TAC, for instance, besides its technical aspect, is an essential tool in the dispute resolution system. The regular meetings of the parties allow the establishment of a discussion forum and help to tackle possible disputes or misunderstandings in due time. TACs are as much an Alternative Dispute Resolution Mechanism as they are a Dispute Prevention Mechanism. They implement a cyclical and informal discussion process, during the whole life of the contract and even in the absence of a dispute. The Dispute Prevention aspect of DRBs is well documented. It is one of the reasons they are so widely

---

33 See for instance, Model Lease Agreement, Article 23.1. (e).

encountered in construction contracts.[34] Besides their high success rate, the implementation of a DRB in a contract leads to a diminution of the number of disputes submitted to it.[35] This decrease mandates considering DRBs as Dispute Prevention Mechanisms, as well as Alternative Dispute Resolution Mechanisms. It is believed that this Dispute Prevention aspect of DRBs occurs due to the fact that the mere existence of the Board encourages the parties to settle their dispute without presenting it to the Board.[36] SED is mostly an ADR mechanism. As opposed to TACs and DRBs, its implementation is simultaneous with the initiation of the dispute. However, the existence of the Expert and his/her professionalism and knowledge in the field, can be a deterrent for the parties in pursuing a further dispute resolution stage, as well as an incentive to amicably settle their dispute.[37] In that respect, SED may be considered as a DPM as well.

2 Interaction between arbitration and the unconventional ADRMs

The use of unconventional ADRMs in oil & gas contracts should not be seen as a threat to arbitration, but rather as a means to solidify it and ensure its position as the preferred forum to solve the disputes of the field. The difficulties arbitration is facing today in relation to costs and – especially – time, could be alleviated by resorting to unconventional ADRMs at the early stages of disputes.

2.1 Unconventional ADRMs as an alternative to arbitration

Unconventional ADRMs have a role to play as pure dispute resolution mechanisms in the oil & gas sector. This specific position is especially true for DRBs and SED. Indeed, both of these mechanisms allow for the resolution of the dispute by one or several experts/technicians, in a limited time period, and as soon as the dispute arises. DRBs present themselves as a good alternative to arbitration for mostly technical disputes. DRBs have a very high success rate, reaching up to more than 90 %.[38] The University of Washington, for instance,

---

34 Kathleen M. J. Harmon, "Construction Conflicts and DRBs", cit., 73.
35 John R. Kohnke, "Dispute Review Boards Rising Star of Construction ADR", *Arbitration Journal* 48, no. 2 (1993), 55.
36 Kathleen M. J. Harmon, "Construction Conflicts and DRBs", cit., 73.
37 Ibid.
38 Colleen A. Libbey, "Working Together while Waltzing in a Mine Field", cit., 844; Christopher T. Horner, "Dispute Review Board Recommendations", cit., 1; Donald Charrett, "The Commercial Value of Dispute Boards", cit., 217–218.

resorted to DRBs for more than 60 projects for an overall amount of more than six billion dollars. Out of these 60 projects, only two disputes required a formal hearing before the board, and there were neither litigation nor arbitration proceedings engaged.[39] The efficiency of DRBs in the sense of absence of further proceedings before national courts or arbitral tribunals can be attributed to different factors. Parties consider recommendations because they trust the impartiality of the Board's members due to their experience and their knowledge of the project.[40] Furthermore, the admission of the recommendation in further proceedings creates an incentive for parties to comply with it, as they expect that judges and arbitrators may follow it.[41] DRBs are a definitive dispute resolution mechanism. Almost all of the disputes referred to a DRB end up being solved without any further proceedings, making DRBs a strong candidate for an alternative to arbitration in the oil & gas industry. DRBs are efficient in terms of their final nature and their time-saving character, but also interesting regarding their cost-saving character as an Alternative Dispute Resolution Mechanism. This cost-saving feature results from the fact that DRBs favour an early dispute resolution, which can be imputed to the recommendations made by the DRB, but also to the mere existence of the Board.[42] Indeed, professionals admit that the existence of a DRB is an incentive for the parties to solve their dispute without its intervention.[43] The cost-saving aspect of DRBs does not cover only financial costs. Due to its effectiveness in solving issues at an early stage, a DRB will also have an impact on the hidden costs of disputes, related to reputation and moral hazard for the contractors.[44] Also, DRBs appear cheaper than mediation. According to the data collected by Dr. Harmon regarding the dispute resolution processes for the Central Artery/Tunnel Project in Boston, the cost of DRB per contract was US$39,130.

---

39 Deborah Bovarnick Mastin, "Government Construction Headaches: Dispute Review Boards to the Rescue", *Public Lawyer* 23, no. 2 (2015), 4.
40 Daniel D. McMillan, "An Owner's Guide", cit., 186.
41 Peter H. J. Chapman, "The Use of Dispute Boards on Major Infrastructure Projects", *Turkish Commercial Law Review* 1, no. 3 (2015), 223.
42 Colleen A. Libbey, "Working Together while Waltzing in a Mine Field", cit., 844.
43 James Denning, "More than an underground success", *Civil Engineering* 63, no. 12 (1993); Harmon, "Effectiveness of DRBs", cit., 676.
44 Peter H. J. Chapman, "The Use of Dispute Boards", cit., 223; Wenxue Lu, Lihan Zhang, and Jing Pan, "Identification and analyses of hidden transaction costs in project dispute resolutions", *International Journal of Project Management* 33, no. 3 (2015), 711–713.

Contrariwise, the mediation conducted under six contracts averaged US$59,024 per contract.[45] The cost of establishing a DRB may sometimes be a deterrent for parties to pursue it. However, it has been demonstrated that DRBs tend to account for 0,1 to 0,25 % of the final contract price.[46] Moreover, for smaller contracts or contracts of lower value (ten million dollars or less), the DRB may be constituted of one member only.[47] Also, the overall cost of the Board can be reduced by spreading the frequency of the meetings.[48] In addition, the same DRB can be established for different contracts and operate project-wise.[49] It stems from the above that DRBs, when considering their effectiveness, definite character and cost-saving aspect, may replace arbitration as an Alternative Dispute Resolution Mechanism. However, as will be explained in the following section, such a replacement is limited. Regarding SED, their implementation in oil & gas contracts is made to avoid arbitration and replace it by a faster and more economical Alternative Dispute Resolution Mechanism. This can be evidenced by the wording of the Greek Model Lease Agreement, in which arbitration is restricted to issues that cannot be solved by SED (outside the scope given to SED by the Lease Agreement).[50] The specific wording is the following: "Any dispute, controversy or claim arising out of or relating to this Agreement [...] which is not to be referred for determination by a Sole Expert under Article 23.1; or has been referred to the Sole Expert whose decision is appealed on a point of law; or if the Lessor has not appointed a Sole Expert [...] shall be finally settled by arbitration."[51] It is clear from this article that arbitration must be seen as a subsidiary to SED, and that arbitration intervenes only when SED failed or could not be actioned under the contract.

---

45 Kathleen M. J. Harmon, "Case study as to the effectiveness of dispute review boards on the central artery/tunnel project", *Journal of Legal Affairs and Dispute Resolution in Engineering and Construction* 1, no. 1 (2008), 25.
46 Charrett, "The Commercial Value of Dispute Boards", 216; Mastin, "Government Construction Headaches", 6; Carol C. Menassa and Feniosky Peña Mora, "Analysis of dispute review boards application in US construction projects from 1975 to 2007", *Journal of Management in Engineering* 26, no. 2 (2009), 66.
47 Mastin, "Government Construction Headaches", cit., 5; Charrett, "The Commercial Value of Dispute Boards", 217.
48 Chapman, "The Use of Dispute Boards", cit., 226.
49 Ibid., p. 219.
50 Model Lease Agreement, Article 23.2 (a).
51 Model Lease Agreement, Article 23.2.

This choice of SED as an alternative to arbitration is well reasoned. Firstly, the decision of the Sole Expert is binding on the parties. Considering SED only as the first step towards arbitration would deprive the whole procedure of its purpose. SED needs to be a final Alternative Dispute Resolution Mechanism. If not, parties will go directly to arbitration and provide expert determination during the arbitral procedure instead, or, in the absence of strict rules of appeal, they would challenge the Sole Expert's decision if they disagreed with it. The SED process needs to be final because it does not share the same characters as DRBs. The Sole Expert is not appointed from the inception of the contract; he/she is sitting alone in the dispute resolution process and is not as accustomed to the parties and the project. For these reasons, and opposed to DRBs, the functioning of SED compels a finality of the decision and requires the procedure to be an alternative to arbitration. Secondly, although the procedure is faster than arbitration, it is still an Alternative Dispute Resolution Mechanism requiring some level of involvement from the parties. They have to agree on the Sole Expert, have to provide submissions, may have to appear to hearings and be ready to disclose documents necessary to the solving of the dispute. In addition, there are some costs attached to a SED. The parties have to pay the Sole Expert and, although the costs are generally shared among the parties, it is an expense compared to a TAC which exists without any costs to be borne by the parties. Moreover, during a SED procedure, parties may wish to be accompanied by a lawyer, who will have to be paid as well. This second reasoning also applies to DRBs. Even if the costs, as demonstrated, are low compared to the overall cost of a project, they remain an investment that is made by the parties, and that has to be protected. Notwithstanding the foregoing, arbitration is not deemed to disappear by the implementation of DRBs or the generalisation of SED. On the contrary, these procedures can still fail, and in these instances, arbitration is a formidable tool and forum for the parties.

2.2 Unconventional ADRMs as a prior step to arbitration

TACs need to be seen as a prior step to arbitration, and not as a proper alternative to it. TACs are a crucial discussion and informal negotiation forum, but they do not provide for definitive measures to settle a dispute that may arise. Consequently, contracts set out subsequent steps to be taken in the event where the parties to the TAC were unable to agree. In the Greek Model Lease Agreement, the TAC must refer any unresolved issue at the end of a meeting to the parties within 15 days, and then parties have to agree within 30 days. After this time, parties are free to refer the dispute to SED or arbitration.[52] TACs are, by their nature, only a first step in the dispute resolution process, and parties should not rely only on this mechanism to solve their disputes. However, the use of a TAC before going to arbitration may force the parties to set out clearly the scope of their dispute, and maybe resolve part of it before going to arbitration. Despite having primarily a definitive character, DRBs, may also be a first step in the dispute resolution process. If a party refuses to abide by the recommendations of the Board, it cannot be forced as the recommendations are non-binding by nature. Parties may, therefore, engage in further proceedings to either obtain the recognition and the application of the recommendation or challenge it.

Arbitral tribunals and litigation at times have been paramount in determining the extent of DRBs' clauses and recommendations. In the case-law saga *CRW Joint Operation v. PT Perusahaan Gas Negara (Persero) TBK*, a binding but not final recommendation had been issued by a DRB (established under a FIDIC DRB clause) against PGN (PT Perusahaan Gas Negara). This decision imposed on PGN to pay a certain amount of money to CRW. CRW (*CRW Joint Operation*) sought its enforcement before two consecutive ICC arbitral tribunals, which was approved by both. The first ICC tribunal rendered a final award, whereas the second issued a partial award.[53] Both arbitrations were followed by courts proceedings before Singaporean courts. These proceedings allowed to refine the link between DRB and arbitration. More interestingly, both jurisdictions upheld certain principles of DRBs. As such, the High Court embodied the principle of "pay now, argue later" upon which DRBs are created.[54] The Court of Appeal recognised the right for a party to seek enforcement of a binding DRB's recommendation before an arbitral tribunal for reasons pertaining to the nature of DRBs. The Court of Appeal recalls that compliance with a DRB's recommendation may be vital for the receiving party to continue arbitral proceedings.[55]

According to the Court of Appeal, parties should be able to immediately seek enforcement and compliance with binding DRB's decisions before arbitral tribunals, without needing the merits of the dispute to be heard first. Indeed,

---

52 Model Lease Agreement, Article 4.10.
53 [2010] SGHC 202; [2011] SGCA 33; [2014] SGHC 146; [2015] SGCA 30.
54 [2014] SGHC 146, para. 157; Yasemin Cetinel, "The Nature of Dispute Board Decisions, with Special Emphasis on the Turkish Law Approach", *Turkish Commercial Law Review* 2, no. 1 (2016), 106.
55 [2015] SGCA 30, para. 73.

an arbitration regarding the non-compliance with the decision will issue an award faster than an arbitral tribunal deciding on the merits of the case.[56] The Court of Appeal of Singapore, therefore, acknowledged the importance in guaranteeing the effectiveness of DRBs. However, it also appreciates the ultimate reliance of DRBs on arbitration in the event of non-compliance of a party with the DRB's decision. Arbitration is, therefore, necessary to DRBs, as it is the only existing mechanism to ensure compliance with recommendations and decisions. Regarding SED and its relationship with arbitration, it can be noted that it is rather limited. The nature of SED is to be a final Alternative Dispute Resolution Mechanism. Arbitration has, therefore, a residual role as an ADRM once a Sole Expert has been involved in the dispute resolution process. For instance, arbitration is often only allowed after SED when there exist errors on points of law. Errors on points of facts cannot be referred to arbitration.[57] But even if the scope of appeal of a SED is limited, arbitration acts as a safeguard to ensure the respect of the procedure.

3 Considering a multi-tiered dispute resolution clause

Multi-tiered dispute resolution clauses, or escalation clauses, are prevalent in long-term contracts such as hydrocarbons exploration and production contracts. They indeed act as a filter, limiting access to arbitration to the most complex disputes.[58] As a consequence, the efficiency of the resolution of disputes is increased, and costs are reduced.[59] Moreover, by creating a sequence of ADRMs from cooperative to adversarial, they allow preserving the relationship between the parties at a dispute stage.[60] The compulsory character of each step will depend on the wording of the clause and on the timeframe provided for each mechanism.[61] Regarding this latter element, parties should rely on objective means of calculation of time, such as days and months. Using

---

56 [2015] SGCA 30, para. 76.
57 Model Lease Agreement, Article 23.2 (b).
58 Craig Tevendale, Hannah Ambrose, and Vanessa Naish, "Multi-Tier Dispute Resolution Clauses and Arbitration", *Turkish Commercial Law Review* 1 (2015), 32–33; Klaus Peter Berger, "Law and practice of escalation clauses", *Arbitration international* 22, no. 1 (2006), 1.
59 Michael Pryles, "Multi-tiered dispute resolution clauses", *Journal of International Arbitration* 18, no. 2 (2001), 159.
60 Klaus Peter Berger, "Law and practice of escalation clauses", 2.
61 Ibid., p. 4.

terms such as 'failure of negotiations' may not have the same meaning for the parties, hence creating another dispute as to whether progression to the next dispute resolution mechanism is possible.[62] Unconventional ADRMs are often found in oil & gas contracts as part of a multi-tiered dispute resolution clause. The contracts of seven different countries (Greece, Cyprus, Lebanon, Syria, Libya, Jordan and Egypt),[63] of the East Mediterranean region were compared as to reach this conclusion. All of these contracts provide for the resolution of disputes by arbitration. All contracts, except the Syrian, Libyan and Egyptian, provide for an escalation clause, by which disputes will be solved through different levels of ADRMs.

The standard dispute resolution clause consists of negotiation or amicable settlement, SED, and finally arbitration. TACs are also implemented in five contracts. However, they are not included as a proper step of the dispute resolution process. They are generally implemented through a specific clause, at the beginning of the contract. It stems from this observation that TACs have yet to be considered as a real ADRM, able to settle technical but also legal disputes at their earliest stage. A possible way to involve TACs in the dispute resolution process would be to link the TAC's procedure to another mechanism, which would then act as an 'appeal' or second-step mechanism.[64] Unlike TACs and SED, none of the contracts provide for DRBs. Their efficiency is recognised, but their insertion in an escalation clause may seem redundant if there already is a TAC or a SED clause included. DRBs and SED are too close in their dispute resolution aspect to be used conjointly. Furthermore, if implemented together, the high success of DRBs would leave a residual role to SED. In this regard, DRBs could be used not as a supplementary step in an escalation clause, but as an alternative to SED at a later stage of the contract. Modulable escalation clauses between the exploration phase and the development/production phase could be envisaged. The development/production stage is indeed construction intensive. In this regard, DRBs could be a great asset in petroleum contracts, as they are a tested and approved method for construction disputes. Resort to arbitration for unresolved issues is always possible, and given the precedents of DRBs, it will be reduced to the most complex issues.

---

62 Pryles, "Multi-tiered dispute resolution clauses", cit., p.160.
63 All of the contracts are accessible and available for download at: https://www.resourcecontracts.org/countries.
64 Model Lease Agreement, Article 4.10.

Conclusions

Unconventional ADRMs may be the way forward in the prevention and resolution of disputes in the oil & gas sector. TACs have already been implemented in several contracts in the East Mediterranean region. Their cyclical and non-adversarial character is crucial in creating and upholding good relationships between the parties to the contract, whilst tackling disagreements and disputes as soon as they arise. Similarly, SED is regularly used as a dispute resolution mechanism, most specifically for technical issues. The procedure is effective in terms of costs and time spent. Moreover, the joint appointment of an expert limits the risks of partiality often encountered with party-appointed experts. DRBs are a specific ADRM, and their implementation is still scarce if not inexistent in oil & gas contracts. They present features from both TACs and SED, and practice, as well as academia, recognise their high efficiency as a dispute prevention and resolution mechanism. Their implementation at the development and production phase of oil & gas contracts could be a good alternative to TACs. Despite their evident benefits for the field with regards to dispute resolution, unconventional ADRMs are not meant to replace arbitration. On the contrary, they should be used conjointly. Arbitration is today the only ADRM internationally recognised and enforceable before national courts. Furthermore, as opposed to TACs and DRBs, the awards are definitely binding on the parties. However, arbitration, in order to be effective, should be a last resort mechanism when all other means have failed. It is hoped that better use of unconventional ADRMs, designed to solve technically challenging disputes, will filter the number of disputes submitted to arbitration.

Dr. Anna Plevri[*]

7 Dispute settlement resolution mechanism in the Greek current exploration licensing rounds

Abstract: This chapter discusses the dispute settlement resolution mechanism in the Greek current exploration licensing rounds and especially arbitration as a resolution method for disputes arising out of the exploration and exploitation of hydrocarbons in the framework of the Greek legal order. The paper presents the latest developments in the Greek framework of exploration licensing rounds and the legislative framework of exploration and exploitation of hydrocarbons in Greece from year 1995 to 2018. Additionally, the calls for international tenders for the right of exploration and exploitation of hydrocarbons in various areas of Greece are discussed among four laws of year 2018 on the ratification of lease agreements between the Greek State and various energy companies on the grant of the right to explore and exploit hydrocarbons in various land or marine areas of Greece.

Keywords: energy disputes, arbitration, exclusive expert, ADR, hydrocarbons, exploration, exploitation, Greece

1 Disputes arising out of the exploration and exploitation of hydrocarbons and arbitration

The 'young' field of energy law is linked to various areas of law such as public international, European, administrative, contractual, commercial, insurance, even criminal law. The energy sector and the sector of natural resources are sectors where conflicts frequently arise at an international level. In order to understand this fact, one should have in mind the economic and energy framework worldwide, which contains economic downfall, increasing lack of energy supplies, plummeting oil prices, rising intra-state energy transit, policies of protecting natural resources, and growing environmental concerns. Having said that, energy-related companies, states, investors and other stakeholders need to take into account the factors of time and cost in order to choose an appropriate, suiting, dispute resolution method. In addition, due to other factors such as the complexity of energy disputes, the length of court proceedings and the

---

[*] Dr. Anna Plevri is Assistant Professor of Private Law and ADR, School of Law, University of Nicosia, Attorney at Law (Greece/Cyprus), Arbitrator and Mediator.

increasing political significance of energy-industry allocation, ADR methods and especially arbitration became a frequent and powerful tool, used by parties internationally in order to protect and safeguard valuable economic interests and rights. The key factors in answering the question of choosing arbitration over litigation in order to resolve energy disputes are the important benefits of confidentiality, neutrality and experience related to the energy sector on behalf of the arbitrator(s). Moreover, energy disputes (and the energy markets in general) have certain characteristics, such as complexity, which requires international energy industry experience, foreign investment expertise and profound technical knowledge.[1] In any case, proficiency in these skills is not usual when it comes to judges and/or jurors. As a result, energy-related companies who are unprepared or not aware of ADR methods are likely to lose market share and face economic losses, in case of the rise of an energy dispute. Under this framework, arbitration is a suitable and well-established method of dealing with conflicts in the energy sector. A valid and binding arbitration clause is therefore needed in parties contracts. Additionally, energy disputes often have a very high monetary value, a strong public interest, and a cross-border character due to the origin of the parties involved. Furthermore, projects for the exploration and exploitation of natural resources require significant financial commitments. This is why the amounts at stake in energy cases are typically among the highest in international arbitration cases. Furthermore, according to the main findings of an initial international report on dispute resolution in the energy sector of the International Center for Energy Arbitration,[2] known as ICEA, presented in May 2015 at the World Forum for Dispute Resolution of the Oil and Gas Industry in Amsterdam, there is a need and strong "support" to resolve energy disputes at an early stage (early dispute settlement), e.g. via mandatory high-level negotiation. The preferred method of dispute resolution was according to above the report arbitration, especially when hybrid arbitration processes (such as med-arb[3]) are taken into account. Mediation has also been well ranked.

---

1 Johan Billiet, "Introductory Remarks", in: *Alternative Dispute Resolution in the Energy Sector*, ed. Association for International Arbitration (Antwerpen/Apeldoorn/Portland: Maklu, 2009), p. 11.
2 "Initial Report on Dispute Resolution in the Energy Sector", *ICEA*, http://www.energyarbitration.org/icea-initial-report-on-dispute-resolution-in-the-energy-sector (accessed 27 September 2019).
3 Regarding "med-arb clauses" see, George Vlavianos and Vasilis Pappas, "Multi-Tier Dispute Resolution Clauses at Jurisdictional Conditions Precedent to Arbitration", in: *The Guide to Energy Arbitrations*, general editor J William Rowley QC, eds. Doak Bishop and Gordon E Kaiser (London: gar, 2017), pp. 217–231.

Regarding the arbitration seat and the factors to be considered when the seat is chosen, the most important factor was related to whether the state of the seat had signed the New York Convention for the Recognition and Enforcement of Arbitral Awards.

The next most important factor of choice was the "reputation" of local courts. The sector of exploration and exploitation of hydrocarbons especially, requires long-term and multilateral agreements,[4] which usually remain in force for several years, so during these contracts, disputes are arising between the parties which can be resolved, and are actually resolved, through international arbitration if they cannot be settled otherwise. In order, of course, for an arbitration process to take place, there is a need for an arbitration clause. Energy projects, by nature, are of great political importance.

The State's role in the exploration and exploitation of hydrocarbons is triple: (a) the State is the one who provides licenses since the ownership of natural resources is in the hands of the State, (b) the State has a legislative role and is responsible for the legal and regulatory framework of this sector and (c) the State also has a contractual role since it is a contractual party via its Hydrocarbons Management Company.

It is also notable that disputes arising out of exploration and exploitation of hydrocarbon agreements are multidimensional. When it comes to methods for the resolution of the above disputes, these agreements usually contain important contractual clauses, (a) "clause for the stabilization, adjustment and modification of the contractual relationship via renegotiation", and (b) "arbitration clause" which often leads to the internationalization of the dispute if it is not already international based on the characteristics of the dispute.[5] The "clause for the stabilization of the contractual relationship" seeks to prevent the State from modifying the legislative and regulatory framework since such a modification would adversely affect the rights and interests of the licensed oil & gas companies as those are protected by the hydrocarbon contract. The success of this clause should not be considered as given because of the fact that several states, especially developing countries, have made significant legislative changes despite the existence of the above contractual clause. In any case, the stabilization clause has been proved as a useful "instrument" for energy

---

4   "Επενδυτικές διαφορές στον τομέα εξερεύνησης και παραγωγής υδρογονανθράκων", https://www.stockwatch.com.cy/el/article/emporika-nea/ependytikes-diafores-ston-tomea-exereynisis-kai-paragogis-ydrogonanthrakon (in Greek; accessed 19 September 2019).

5   Regarding these contractual clauses, see Loukas Mistelis, "Contractual Mechanisms for stability in energy contracts", in: *International Arbitration in the Energy Sector*, ed. Maxi Scherer (Oxford: Oxford University Press, 2018), pp. 153–174.

companies in the process of resolving investment disputes by international arbitration tribunals, mainly according to *Bilateral Investment Agreements (BIT's)*,[6] and *Multilateral Investment Agreements (MIT's)*,[7] such as the *Energy Charter Treaty (ECT)*[8]. It is further true, that the "internationalization of the

---

6   A BIT provides investors of one country with the right to bring arbitration claims against another country in certain circumstances.
7   See among others, Chin Leng Lim, Jean Ho and Martins Paparinskis, "International Investment Law and Arbitration, Commentary, Awards and other Materials" (Cambridge: Cambridge University Press, 2018); Aloysious Gng, "The concept of soft law investment arbitration", in: *Alternative Dispute Resolution in the Energy Sector*, ed. Association for International Arbitration (AIA) (Antwerpen/Apelddorn/Portland: Maklu, 2009), pp. 95–114; Hege Elisabeth Kjos, "Applicable Law in Investor-State Arbitration, The Interplay between National and International Law" (Oxford: Oxford University Press, 2013); Anna Plevri, "Arbitration in the energy sector (Greek and European institutional framework)" (Athens-Thessaloniki: Sakkoulas Publications, 2018), pp. 15–44 (in Greek).
8   On the content of ECT, see "The Energy Charter Treaty and Related Documents, A Legal Framework For International Cooperation (2004)", *Energy Charter Secretariat*, http://www.encharter.org/file/admin/userupload/document.Eu.pdf (accessed 27 September 2019); and among other authors, Maxi Scherer (ed.), in: *International Arbitration in the Energy Sector* (Oxford: Oxford University Press, 2018); Cyrus Benson, Charline Yim and Victoria Orlowski, "The Energy Charter Treaty", in: *The Guide to Energy Arbitrations*, general editor J William Rowley QC, eds. Doak Bishop and Gordon E Kaiser (London: gar, 2017), pp. 30–46; Thomas Roe and Matthew Happold, Consultant Editor James Dingemans QC, "Settlement of Investment Disputes under the Energy Charter Treaty" (Cambridge: Cambridge University Press, 2011), p. 18; Clarisse Ribeiro, "Investment Arbitration and the Energy Charter Treaty" (Huntington: Jurisnet, 2006), pp. 1–47; Matthew Slater, "The Energy Charter Treaty: A brief introduction to its scope an initial arbitral awards", in: *Alternative Dispute Resolution in the Energy Sector*, ed. Association for International Arbitration (AIA) (Antwerpen/Apelddorn/Portland: Maklu, 2009), pp. 15–54; "Annex 2, ECT, Part V, Dispute Settlement", in: *Alternative Dispute Resolution in the Energy Sector*, ed. Association for International Arbitration (AIA) (Antwerpen/Apelddorn/Portland: Maklu, 2009), pp. 137–145; Greece has signed the ECT on 17.12.1994 and ratified it on 16.07.1997. Greece has also signed the International Energy Charter Treaty on 20.05.2015. International Charter Treaty is a declaration of political intention aiming at strengthening energy cooperation between the signatory states and does not bear any legally binding obligation or financial commitment. The International Energy Charter promotes mutually beneficial energy cooperation among nations for the sake of energy security and sustainability. See more at "The International Energy Charter Overview", *Energy Charter Treaty*,

dispute" (by including an international arbitration clause in an agreement of exploration and exploitation of hydrocarbons) creates a "security framework" for the oil & gas companies, since in the event of the termination or breach of the contract by the state, the dispute is resolved by an international arbitration tribunal, which applies principles of international law in addition to the applicable rules of law according to the contract. In addition, investment agreements provide additional protection, since a dispute arising from a "hydrocarbon contract" may result in an investment dispute, because an energy company as a contractual party investor, may claim a simultaneous breach of the applicable investment agreement.

Compared to the "stabilization clause" of a contract, the "renegotiation clause" aims to maintain and improve the contractual relationship. In any case, both clauses are important "instruments" in order to enhance the stability and the effort of maintaining the contractual relationship. This applies, especially because there are significant exogenous factors and possible changes in the field of exploration and exploitation of hydrocarbons which may have a profound effect on the rights of the parties. In contrast to the evolving nature of the energy industry, is the investors' need for stability. Investors are keen to keep the legislative, regulatory, political and economic framework of their energy project as stable as possible. Moreover, an important common feature of energy projects is the interconnected nature of commercial arbitration and investor-state arbitration.[9]

A single energy project might parallelly trigger different types of arbitrations. For instance, if a host state wishes to involve a foreign investor in the exploration and production of hydrocarbons, there is often an agreement between the host state (or a state-related entity) and the investor, in the form of a concession or production sharing agreement (PSA) to regulate the terms of the investment. It is common that the investor is a joint venture of several companies, and their legal relations with each other are regulated in JOA (Joint Operations Agreement). The JOA might lead to a commercial arbitration between the joint ventures, while an investor-state arbitration might be initiated in their relations with the host state. The latter could be either contract-based (e.g. an

---

http://www.energycharter.org/process/international-energy-charter-2015/overview/ (accessed 27 September 2019).

9 See Maxi Scherer (ed.), "Part I: Commercial Arbitration in the Energy Sector", in: *International Arbitration in the Energy Sector* (Oxford: Oxford University Press, 2018), pp. 21–149 and Maxi Scherer, "Part II: Investor-State Arbitration in the Energy Sector", in: *International Arbitration in the Energy Sector* (Oxford: Oxford University Press, 2018), pp. 153–275.

arbitration agreement in the Production Sharing Agreement (PSA) or treaty-based (e.g. under an applicable BIT). Of course, the significant and frequent exogenous changes in the field of hydrocarbons have a severe effect which changes the bargaining position and power of the parties. This inevitably "gives birth" to a dispute. Therefore, it is imperative when negotiating, drawing up and signing an agreement of exploration and exploitation of hydrocarbons or any energy contract, to seek the incorporation of a "multi-level strategy" a "multi-tier/multi-level dispute resolution clause" (renegotiation – mediation – arbitration), which includes a comprehensive assessment and effective use of out of court dispute resolution and settlement methods.

2 Latest developments in the Greek framework of exploration licensing rounds

It is widely accepted that gas will play a central role in the energy mix of Greece in the coming years and this makes imperative the need of searching for new sources especially in view of the goal of ensuring the country's energy security. The last two years, one can notice intense and important movements in the sector of hydrocarbon exploration in Greece and subsequently in the licensing rounds.[10] More specifically, after April 2019,[11] four hydrocarbon exploration and exploitation rights agreements were signed[12] with consortia of companies and are expected to be ratified by the Parliament. The ratification of those agreements was also announced by the Energy Minister, Mr. K. Hadjidakis during his contacts with the US Secretary of State F. Fannon before August 15, 2019. The Ministry of Energy is in cooperation with the Hellenic Hydrocarbons Management Company in these agreements. The Hellenic Hydrocarbons Management Company constitutes the "technocratic arm" of the Greek State in

---

10 For historical record of hydrocarbon exploration in Greece see http://www.ypeka.gr/Default.aspx?tabid=765&locale=en-US&language=el-GR (in Greek; accessed 17 September 2019).

11 "Εντός του Αυγούστου στη Βουλή προς κύρωση οι συμβάσεις παραχώρησης υδρογονανθράκων για Κρήτη – Ιόνιο – Κυπαρισσιακό", *energypress*, www.energypress.gr/news/entos-toy-aygoystoy-sti-voyli-pros-kyrosi-oi-symvaseis-parahorisis-ydrogonanthrakon-gia-kriti (in Greek; accessed 27 August 2019).

12 Two of them were signed in April 2019 and two in June 2019, see "Υπογράφονται οι συμβάσεις για έρευνες υδρογονανθράκων στην Κρήτη", *LIFO*, www.lifo.gr/now/greece/242493/ypografontai-oi-symvaseis-gia-ereynes-ydrogonanthrakon-stin-kriti-ti-perilamvanoyn (in Greek; accessed 17 September 2019).

this field. In all of these projects major international energy companies and/or consortia are involved: Two of the above agreements are concerning areas in the south and southwest of Crete,[13] (*consortium of ELPE (ELPE Upstream) (Greek Petroleum), French Total & the American Exxon Mobil*). In the Ionian Sea, the agreement has been signed with *ELPE* and the *Spanish Repsol Exploration SA*, while in the Gulf of Kyparissia (Block 10) the agreement has been signed with (*ELPE*) (*Greek Petroleum*).

The ratification of the above four agreements is, certainly, of great importance since it is directly connected with the implementation of significant investments in the energy sector in Greece. This significant development is part of the country's strategic pursuit of obtaining an international presence in the critical sector of energy.[14] As noted,[15] the most real immediate benefit comes from the message being conveyed to the international investment community, especially that of the energy sector, that Greece is gradually regaining the position of a secure investment destination and that it has an overall strategy of long-term targeting. Naturally, positive results are also expected for the Greek economy through the activation of the consortium on these plots and the benefits arising from the development of the economic activity resulting from the first phase of the exploration work to be carried out. The practical significance of the above agreements is inextricably linked to the outcome of exploration work and whether there are sufficient quantities of utilizable and economically exploitable energy resources to carry out the third phase of the contractual stages of drilling.

---

13 Areas of great geopolitical importance in the Greek territory.
14 Antonis Metaxas, "Significance of the signing of hydrocarbon contracts in the maritime region west and southwest of Crete", *energypress*, 1 July 2019, https://energypress.gr/news/i-simasia-tis-ypografis-ton-symvaseon-ydrogonanthrakon-sti-thalassia-periohi -dytika-kai (accessed 19 September 2019).
15 Ibid.

3 Legislative framework of exploration and exploitation of hydrocarbons in Greece

3.1 Law 2289/1995 (Hydrocarbons Law) as amended by Law 4001/2011

Law 2289/1995 on the Search, Exploration and Exploitation of Hydrocarbons (and other provisions),[16] reformed the institutional framework for licensing,[17] and has incorporated into the Greek law, the European Directive 94/22/EC of 30 May 1994 on the conditions for the licensing of exploration and exploitation of hydrocarbons.[18] The above governing law in this field was amended by Law 4001/2011, which modernized the regulatory framework for hydrocarbons,[19] drilling and exploration in Greece, so many provisions of Law 2289/1995 were replaced. According to Law 2289/1995, as amended by Law 4001/2011, the

---

16 See in detail on the "Hydrocarbons Law": Anna Plevri, "Arbitration in the energy sector (Greek and European institutional framework)" (Athens-Thessaloniki: Sakkoulas Publications, 2018), pp. 57–66; Theodoros Fortsakis and Nikolaos Farantouris, "Energy Law" (Athens: Nomiki Bibliothiki, 2016), pp. 284–288 (in Greek); Theodoros Fortsakis, "Energy Law" (Athens: Sakkoulas Publications, 2009), pp. 275–278, (in Greek).

17 The first law related to exploration of hydrocarbons in Greece was law 468/1976.

18 See Timoleon Kosmides, "The License Agreements of Exploration and Exploitation of Hydrocarbons (on the Basis Law No 2289/1995) – A First Approach", *Nomiko Vima* (2018), 788 et seq. (in Greek); Timoleon Kosmides, "The Agreements on rights of exploration and exploitation of hydrocarbons", in: *Energy, Networks and Infrastructure*, ed. Nikolaos Farantouris (Athens: Nomiki Bibliothiki, 2014), pp. 421 et seq. (in Greek); see also the Directive 2013/30/EU of the European Parliament and of the Council of 12 June 2013 on safety of offshore oil and gas operations.

19 According to Article 1 § 1 of Law 2289/1995, the definition of hydrocarbons is as follows: "Hydrocarbons: Any kind of mineral oil in solid, liquid or gaseous condition and in particular mineral crude oil or natural gasoline, natural hydrocarbon gases and any minerals or substances extracted with them." Moreover, according to Article 1 § 3 of Law 2289/1995, the "search for hydrocarbons" is defined as the attempt to locate hydrocarbons in a particular area by any convenient method other than drilling, and to the provisions of paragraph 4 of the same Article the "exploration of hydrocarbons" is defined as the exploration on the discovery of hydrocarbon deposits by any convenient method, as well as by drilling. In addition, "exploitation of hydrocarbons", according to Article 1 § 5 of Law 2289/1995, is the extraction of hydrocarbons, their possible treatment in order to become marketable (refining is not included here), their storage and transportation as well as the storage and transportation of their by-products to loading facilities for further disposal.

right for exploration and exploitation is granted by the Greek State (which is a party in this concession agreement) either by signing an agreement of lease or contract of distribution of the production.[20] In both cases, the contract provides for two stages, namely that of exploration and that of exploitation.

Regarding the methods of resolution of disputes arising out of exploration or exploitation of hydrocarbons, article 10 § 12 of Law 2289/1995 (as applicable to the present day) provides that disputes arising out of these contracts are resolved by the Administrative Court of Appeal of Athens since they are considered as administrative disputes regarding the substance of the case. Provisions of Law 1406/1983 are also applicable in this case. The Greek State is a contracting member as a public authority in the above agreements thus the disputes arising out of these agreements belong to the sphere of public law and accordingly are to be heard by an administrative court. Naturally, the inevitable question lies over whether or not an administrative court with a general competence – which is not an energy court – is an appropriate forum for the effective resolution of energy disputes.

Moreover, article 10 § 13 of Law 2289/1995,[21] provides[22] that any dispute under the provisions of this law, contractual (the Greek State is a contracting party) or tortious, is resolved/settled by an arbitral tribunal according to the international commercial arbitration (Greek) Law 2735/1999 or another internationally recognized arbitration system, such as ICC, LCIA, Arbitration Institute of the Stockholm Chamber of Commerce,[23] thus excluding the competence of the Greek courts. The arbitral tribunal shall consist of two arbitrators appointed by the parties and one arbitrator (chair) appointed by the arbitrators. The seat of arbitration is going to be in Athens, the language of proceedings will be Greek, and the applicable law will be Greek. The above provision, therefore, allows the parties to seek resolution of their disputes

---

20 For these kind of energy contracts, see Timoleon Kosmides, in: *Energy, Networks and Infrastructure*, pp. 427–445, where the author argues that the "contract of distribution of the production" is more advantageous to the Greek State instead of the "lease contract". In a lease contract there are provisions not only for the completion of the research project but for the subsequent exploitation of the hydrocarbons.
21 As amended by Article 162 § 5 of Law 4001/2011.
22 The content of Article 10, paragraph 13 of Law 2289/1995, is wider than that of paragraph 12 of the same Article which regards only disputes arising out of the performance of contracts regarding exploration and exploitation of hydrocarbons (according to the wording of paragraph 12).
23 The reference to the above internationally recognized arbitration centres is just indicative.

which have arisen or may arise from a contractual or non-contractual relationship to an arbitral tribunal, thereby excluding the jurisdiction of the Administrative Court of Appeal, in accordance with the standard of the international practice when it comes to "investment agreements". In order to materialize this ability, the parties need an arbitration clause as defined in Article 7 § 2 and § 3 of Law 2735/1999, e.g. by incorporating an arbitration clause in their contract/agreement without prejudice to *jus cogens* provisions of Article 10 § 13 (b) and (c). It should be noted that this provision would be more flexible, thus more "friendly" to the contracting parties in agreements of exploration and exploitation of hydrocarbons which are usually foreign companies or joint ventures if it would provide an option of choice at least regarding the language of the proceedings. It is, therefore, reasonable that article 10 § 13 of Law 2289/1995 intents the conduct of arbitration proceedings in Greece with the participation of arbitrators and attorneys-at-law who know both the Greek language and Greek law. The seat of arbitration, which means the seat of the arbitral tribunal, further determines the application of the provisions of the Greek Code of Civil Procedure or those of Law 2735/99 on the issue of the annulment of the arbitral award in international commercial arbitration in Greece.

In conclusion, the above legislative provisions regard international arbitration for either contractual or tortious disputes,[24] arising out of activities and contracts which are included in Law 2289/1995 on exploration and exploitation of hydrocarbons, meaning agreements of lease or contracts of distribution of the production in which the Greek State is the contracting party.[25] This provision

---

24 As it is commonly accepted in Greek law, disputes which have been arisen or may arise out of a legal relationship, contractual or not may be resolved through arbitration if there is an arbitration clause signed by the parties, see indicatively, Supreme Court 102/2012, *Nomiko Vima* 2012, 1760 (in Greek). The arbitration clause may, therefore, validly include claims arising immediately out of the underlying contract or claims arising out of tort or future disputes, provided that they refer to the underlying legal relationship, Supreme Court 2004/2007; Supreme Court 1737/2009; Supreme Court 506/2010, NOMOS law database; Supreme Court 1219/2014, Archives of Private Law 2015, 130 (in Greek).
25 According to Article 49 of the Introductory Law of the Greek Code of Civil Procedure, in order for the Greek State to validly conclude an arbitration agreement, it has to be in writing and after an opinion of Plenary of the Legal Council (of the State), a decision of the Minister of Finance and the Minister responsible. In the same way, the Greek State can appoint its arbitrators. The time limits of Articles 873 to 876 of the Greek Code of Civil Procedure shall be one month in case of arbitration agreements concluded by the Greek State. Article 6 § 3α of Law 3086/2002 is also applicable (as

of the law refers to a possibility for the parties and not a mandatory arbitration which would not be compatible with the provisions of the Greek Constitution.

3.2 Presidential decree No 127/1996 and Draft Model Lease Agreement

In accordance with the provisions of Law 2289/1995, the presidential decree No 127/1996 was issued, which regulates the conditions for drawing up lease agreements for exploration and exploitation of hydrocarbons, (Article 2 § 10 (a) of Law 2289/1995). Article 11 of the above presidential decree states that a lease agreement may provide that disputes may be resolved through arbitration in accordance with the provisions of Article 10 § 13 of Law 2289/1995. Moreover, as provided by the legislation (Articles 145–164 of Law 4001/2011 (in particular articles 154 § 2 and 153 § 3), presidential decree 127/1996 and article 20 of Law 3587/2007) on January 2012, a decision was issued by the Deputy Minister of Environment, Energy and Climate Change (open door invitation[26]) entitled: "International Public Open Invitation of interest in granting (by the Greek State) the right to exploration and exploitation of hydrocarbons in the marine areas of 'Patraikos Gulf (West)' and 'Katakolo' and in the land area 'Ioannina'". In this context, the Ministry of Energy, Environment and Climate Change published (in English) the draft of the relevant lease agreement to be concluded between the parties (Draft Model Lease Agreement).[27] Article 24 of this draft, under the title 'Sole Expert Determination and Settlement of Dispute', provides (in particular from paragraphs 3–9) in relation to the out-of-court resolution of disputes that: "24.3 Any dispute, controversy or claim arising out of or relating to the Agreement, or breach, termination or invalidity hereof, between the Lessor and the Lessee which: (a) is not to be referred to for determination by a Sole Expert under Article 24.2[28]; or (b) has

---

amended by Article 103 of Law 4139/2013) and provides that in arbitration proceedings (both domestic and international) relating to contracts drawn up with the Greek State or State-owned public or private legal entities established and operating in order to serve the public interest, the Greek State can appoint as an arbitrator a senior judge or a senior officer from the Legal Counsel of the State or a University Professor or an Attorney at Law at the Supreme Court.

26 See www.ypeka.gr/LinkClick.aspx?fileticket=79q2OpjZIKw%3D&tabid=765 (accessed 19 September 2019).
27 See www.ypeka.gr/LinkClick.aspx?fileticket=ydsJhXqPPT8 %3D&tabid=766 (accessed 19 September 2019).
28 "24.2 Where it is stipulated in this Agreement that any difference between the Parties or any inability or failure by the Parties to agree on any matter (a "Dispute") shall be

referred for determination to a Sole Expert, unless the Parties agree otherwise the following shall apply: (a) The Sole Expert shall be selected by the Lessor in accordance with Articles 2.1 and 2.2 of the Presidential Decree within fifteen (15) days (the "Election Period") from submission of a written notification by a Party (the "Initiating Party") to the other (the "Receiving Party") of its intention to refer a Dispute for determination to a Sole Expert and nominating the Sole Expert. (b) Upon a Sole Expert being selected under the foregoing provisions of this Article, the Lessor shall forthwith notify this Sole Expert of its selection and shall request it to state within 5 days (the "Acceptance Period") whether or not it is willing and able to accept the appointment. If such Sole Expert shall be either unwilling or unable to accept such appointment, or shall not have accepted (the "Disqualified Expert") within the Acceptance Period then the Lessor shall select an alternative a Sole Expert in accordance with Articles 2.1 and 2.2 of the Presidential Decree within 5 days following the end of the Acceptance Period. (c) For the purposes of determination by the Sole Expert of the Dispute, each Party shall submit to the Sole Expert within ten (10) days (the "Submissions Period") following the Sole Expert's acceptance of appointment to both Parties: (i) a description of the Dispute; (ii) a statement of its position; (iii) any records supporting its position. The Sole Expert shall consider, in its absolute discretion, any additional information submitted by either Party and any other procedural matters not specifically addressed herein. (d) The terms of reference upon which the Sole Expert shall seek to resolve a Dispute shall be agreed between the Parties. The parameters within which the Sole Expert shall make its determination shall be strictly within the terms of reference, provided that if the Parties fail to agree terms of reference, the Sole Expert shall consider the terms of reference proposed by both Parties and decide upon its own (to which the Parties shall be bound). The Sole Expert shall make its determination in writing and notify the Parties of such determination. (e) The Sole Expert's determination shall be conclusive and binding on the Parties and shall be delivered within fifteen (15) days following the end of the Submissions Period. The decision of the Sole Expert may be referred to arbitration by way of appeal on point of law, but not on point of fact. (f) If the Sole Expert dies or becomes unwilling or incapable of acting, or does not deliver the determination within the time required by this clause then: (i) the Lessor shall promptly select a replacement Sole Expert; and (ii) this Article shall apply to the new Sole Expert as if he were the first Sole Expert appointed. (g) The language to be used for the purposes of Sole Expert Determination shall be English. (h) The costs of engaging the Sole Expert shall be borne equally by the Lessor, on the one hand, and the Lessee, on the other. Each Party shall bear its own costs in preparing materials for and making presentations to, the Sole Expert. (i) Each Party shall act reasonably and co-operate to give effect to the provisions of this clause and otherwise do nothing to hinder or prevent the Sole Expert from reaching his determination. (j) The Sole Expert shall be the person referred to in Articles 2.1 and 2.2 of the Presidential Decree. (k) If the Lessor fails to appoint a Sole Expert (or, as the case may be a replacement Sole Expert) within the time limits prescribed by this Article, then the Lessee shall be entitled to refer the relevant dispute, controversy or claim to arbitration in accordance with the following provisions of this Article".

been referred to the Sole Expert whose decision is appealed on a point of law; or (c) if the Lessor has not appointed a Sole Expert (or, as the case may be a replacement Sole Expert) within the time limits prescribed by this Article, shall be finally settled by arbitration. 24.4 The place of arbitration shall be Athens, Greece.[29] 24.5 The number of arbitrators shall be three and they shall be appointed in accordance with the provisions of paragraph 13 of Article 10 of the Hydrocarbons Law.[30] 24.6 The arbitration shall be conducted in accordance with the Rules of Arbitration of the International Chamber of Commerce (in force from time to time) to the extent that there is no conflict between any of those rules and the provisions of this Agreement and Greek Law. In the event of any such conflict, the provisions of this Agreement and Greek Law shall prevail. 24.7 The language to be used in the arbitral proceedings shall be Greek unless the Parties agree otherwise. 24.8 A decision of the majority of the arbitrators shall be final and binding upon the Parties and the award rendered shall be final and conclusive. Judgment on the award rendered may be entered in any court for a juridical acceptance and for enforcement, as the case may be. 24.9 During the period of any arbitration the time limits, set for the fulfilment by either Party or those contractual obligations under this Agreement which are the subject of such arbitration, shall be suspended for a time equivalent to the period of such arbitration."

4 Calls for international tenders for the right of exploration and exploitation of hydrocarbons in areas of Greece[31]

Pursuant to Greek law,[32] and after applications by specific companies to grant them rights of exploration and exploitation of hydrocarbons, a ministerial decision accepted those applications and the procedure referred in Article 2 § 17 case (b) of the Hydrocarbon Law was initiated. By a Declaration by the Hellenic

---

29 If Athens is the seat of the arbitration, Greek courts will have jurisdiction on procedures of setting aside the arbitral award.
30 Article 10 § 13 Law 2289/1995.
31 An agreement between the Greek State and various companies regarding the exploration and exploitation of hydrocarbons in the sea area of the Thracian Sea dated 14.06.1975 was ratified by Law 98/1975 and included an arbitration clause for the settlement of related disputes (Article 33 of Law 98/1975).
32 Articles 145–164 of Law 4001/2011, Presidential Decree 14/2012 on setting up of the Hellenic Hydrocarbon Resources Management S.A., Article 2 § 17 Case B of the Hydrocarbons Law (2289/1995), Article 148 of the Legislative Decree 210/1973 as it is currently in force, the presidential decree 127/1996 and Article 20 of Law 3587/2007.

Hydrocarbon Company (No 745/18.07.2017) and a decision by the Ministry of Energy (No 14128/4286/27.07.2017) the "Lease Agreement" was selected as a type of contract for the grant of the right to explore and exploit hydrocarbons. Furthermore, a decision of the Ministry of Energy (No 14325/4348/2017) was issued announcing an international call for tender for the exploration and exploitation of hydrocarbons in the marine areas of "West Crete" and "Southwest Crete". According to the above decision the interested parties were invited to submit tenders for the above areas,[33] in the context of the continuing need for continuous, thorough, effective and safe research for the detection of oil and gas in the Greek territory.

Additionally, in accordance to the same legislation and after an application by a specific company for the granting of rights of exploration and exploitation of hydrocarbons, a ministerial decision accepted that application, and the procedure referred in Article 2 § 17 case (b) of the Hydrocarbon Law was initiated. By a Declaration by the Hellenic Hydrocarbon Company (No 743/18.07.2017) and a decision by the Ministry of Energy (No 13967/4241/26.07.2017) the "Lease Agreement" was also selected as a type of contract for the grant of the right to explore and exploit hydrocarbons. Moreover, a decision by the Ministry of Energy (No 14322/4347/2017) was issued announcing an international call for tender for the exploration and exploitation of hydrocarbons in the marine area of "West Greece" (*Ionian*) and the interested parties were invited to submit tenders for the above area.[34] Subsequently, four ratifying laws of the respective lease agreements for the concession of rights of exploration and exploitation of hydrocarbons were concluded in 2018 between the Greek State and energy companies, with respect to four different regions in Greece and more specifically in the land area of north-western Peloponnese, the land area of Arta–Preveza, the marine area of the Ionian Sea and the land area of Aitoloakarnania. The area of Crete is not included in any of these lease agreements and their respective ratification laws.

All of the above four agreements and their respective ratification laws, which constitute ad hoc legislation were designed to resolve energy disputes, contain detailed provisions regarding the resolution of disputes which may arise. It is worthy to note that in three of these laws (Law 4524/2018 (land area of

---

33 In accordance to the procedure referred in Article 2 § 17 Case (b) of Hydrocarbons Law.
34 In accordance to the procedure referred in Article 2 § 17 Case (b) of Hydrocarbons Law. See in detail, Anna Plevri, "Arbitration in the energy sector (Greek and European institutional framework)", pp. 66–68 (in Greek).

Aitoloakarnania), 4526/2018 (land area of Arta–Preveza) and 4527/2018 (land area of Northwest Peloponnese), the relevant term (Article 23), has an identical wording, whereas, in Law 4525/2018 (marine area 2 of the Ionian Sea), the term in question contains different provisions as to how the dispute can be resolved by a "sole expert" before arbitration and how the "sole expert" is chosen. In addition, Article 23 of Law 4525/2018 also provides for mediation as an additional ADR method, in accordance with the ICC mediation rules (clause 23.12) if the parties wish so.

5 Laws 4524/2018, 4525/2018 4526/2018 and 4527/2018 on the ratification of lease agreements between the Greek State and various energy companies on the grant of the right to explore and exploit hydrocarbons in various lands or marine areas in Greece

On the ratification of lease agreements between the Greek State and various energy companies on the grant of the right to explore and exploit hydrocarbons in various lands or marine areas of Greece, Law 4524/2018,[35] Law 4525/2018,[36] Law 4526/2018,[37] and Law 4527/2018,[38] are shortly presented here. Law 4524/2018,[39] ratified the lease agreement between the Greek State (the Hellenic Republic) and the energy company "ENERGEAN OIL & GAS-AEGEAN ENERGY S.A." granting the right to explore and exploit hydrocarbons in the land area of Aitoloakarnania. Law 4525/2018,[40] ratified the lease agreement between the Greek State (the Hellenic Republic) and energy companies "Total E&P Greece B.v", "Edison International S.p.A." and "Hellenic Petroleum S.A." granting the right to explore and exploit hydrocarbons in the marine area 2 of the Ionian sea. Law 4526/2018,[41] ratified the lease agreement between the Greek State (the Hellenic Republic) and the energy company "Hellenic Petroleum S.A." granting the right to explore and exploit hydrocarbons in the land area of "Arta-Preveza". Law 4527/2018,[42] ratified the lease agreement between

---

35 Ibid., pp. 69–72.
36 Ibid., pp. 72–77.
37 Ibid., pp. 77–81.
38 Ibid., pp. 81–84.
39 Government's Gazette A 46/ 15.03.2018. Law 4524/2018 contains 36 Articles.
40 Government's Gazette A 47/ 15.03.2018. Law 4525/2018 contains 36 Articles.
41 Government's Gazette A 48/ 15.03.2018. Law 4526/2018 contains 35 Articles.
42 Government's Gazette A 49/16.03.2018. Law 4527/2018 contains 36 Articles.

the Greek State (the Hellenic Republic) and the energy company "Hellenic Petroleum S.A." granting the right to explore and exploit hydrocarbons in the land area of "Northwest Peloponnese". The subject of all the above four agreements/contracts is defined in the first article of each respective law. These contracts constitute lease agreements, under which the Greek State, (*the lessor*) (according to Article 2 § 10 of the Hydrocarbons Law), assigns to *the lessee* exclusive rights to carry out (in the duration of the contract) oil operations in the contract area under contractual terms and conditions and according to the provisions of the law. The methods to resolve disputes between the parties or any failure of the parties to agree on any matter regarding the interpretation or application of any provision of the contract are described in Article 23 of Laws 4524/2018, 4525/2018, 4526/2018 and 4527/2018 entitled "Resolution/Settlement of Disputes".[43] Article 23 provides for three consecutive steps or methods of out-of-court dispute resolution. The last one is arbitration. More specifically, these methods are i) Friendly settlement/Amicable Resolution, ii) Sole Expert Determination,[44] iii) Arbitration. Moreover, Article 31 of Laws 4524/2018, 4525/2018, 4526/2018 and 4527/2018 (identical in these four laws) refers to the law applicable to the contract, (which was signed in both the Greek and English languages), however, the Greek text is prevailing over any ambiguities. The ratified contract was stipulated that it would be governed and interpreted by Greek law. The third paragraph of Article 31 is noteworthy. It states that: "No article of this contract may derogate or oblige the State to derogate from any provision contained in EU Treaties, including for the avoidance of doubt and of EU laws which are adopted in implementation of the Treaties of the EU."

5.1 Resolution of disputes

The resolution of disputes as provided in Article 23 of each of the Laws 4524/2018, 4526/2018 and 4527/2018 (not Law 4525/2018) is summarized in

---

43 See Anna Plevri, "Arbitration in the energy sector (Greek and European institutional framework)" (Athens-Thessaloniki: Sakkoulas Publications, 2018), pp. 84–90 (in Greek).

44 According to Article one and preamble of Laws 4524/2018, 4525/2018 the "Exclusive Expert" means in the context of this law, a member of (a) the Energy Institute of London, (b) the American Petroleum Institute or (c) the French Institute of Petroleum (IFP Energies nouvelles), provided that in the event of a conflict of interest with all of the aforementioned institutes, the Lessor has the right to appoint an independent recognized oil institute of another country of the European Union in which hydrocarbons are produced.

three steps where Stage 1 refers to Friendly Settlement/Amicable Resolution, Stage 2 to Sole Expert Determination and Stage 3 to Referral to arbitration. On Stage 1 – Friendly Settlement/Amicable Resolution, initially, parties are encouraged to attempt an "amicable resolution" of their dispute through discussions for a period no longer than thirty days after receiving a notice (coming from the other party) for the existence of the dispute. In case of failure to reach an amicable settlement within the above period, the contract provides for the dispute to be referred for determination by the "Exclusive Expert", not for all matters, meaning not for any dispute arising out of the contract but for matters which under the contract may be referred to the "Exclusive Expert". On Stage 2 – Sole Expert Determination, there are detailed provisions for determination of the dispute by the "Sole/Exclusive Expert", who has to be a member of specific international institutes for energy and oil (as already mentioned). In my view, the law deliberately uses the term "determination" at this point and not that of "resolution of the dispute by the Exclusive Expert". The Sole/Exclusive Expert is appointed by the Lessor (announced to the Lessee) and not agreed by the parties. Both parties shall equally, however, bear the cost of the 'Exclusive Expert'. The "Sole/Exclusive Expert" determines his/her "decision" (this term is used by the above law) based on criteria which strictly fall within the terms of reference agreed between the parties or if there are no agreed terms of reference, the 'Sole/Exclusive Expert' shall compulsorily examine the terms of reference proposed by each party and shall make a binding "decision" on its own, "written" in English, irrespective of his/her nationality. In making that "decision", the "Exclusive Expert" may, at his/her sole discretion, take into account any supplementary information submitted by each party and/or any other procedural matters which are not expressly mentioned in the provisions of the law. In accordance with the aforementioned legislation, this "decision" of the "Exclusive Expert" is made within a short period of 30 days after the end of the submission period and is final and binding on the parties with the exception of cases of fraud or obvious error (without however stating in the law what constitutes a manifested error), so in that case, it will not be a "binding decision". On Stage 3 – Referral to arbitration, the "decision"/determination of the "Exclusive Expert" may be referred to arbitration by way of an appeal on a point of law rather than a factual issue/matter. This provision of the law permits the safe conclusion that the "Exclusive Expert" determines only factual issues and not points of law. At this point, one could argue that the use (by law) of the term "decision" in order to describe the determination by the "Exclusive Expert" is not a successful one. The determination of the "Exclusive Expert" is then brought before an arbitral tribunal, after referral of the dispute to arbitration on a point of law. More specifically, the dispute may be referred to arbitration (for a final resolution) if the Lessor has not appointed a Sole Expert (or, as the case may be, a replacement of Sole Expert) within the time limits provided by the law. In the above cases, the Lessee refers the dispute to arbitration in accordance with the provisions of Article 23.

According to the provisions of the above four laws (Article 23.3) the final resolution of the dispute through arbitration is possible for any dispute or claim arising out of, or in connection with, the lease agreement/contract, or any breach, termination or invalidity of this contract, which either is not a dispute referred to the sole discretion of the 'Expert' (under Article 23.1) or has been referred to the Sole Expert (in accordance with Article 23.2 of the contract) but his/her determination/decision has been appealed on point of law, either the lessor has not appointed an "Exclusive Expert" or, as the case may be, a replacement Sole Expert within the statutory deadline. The seat of arbitration shall be in Athens (Article 23.5), so the *lex arbitri* shall be Greek law, and the arbitral tribunal shall be composed of three members in accordance with Article 10 § 13 of Hydrocarbons Law. Arbitration will be conducted in accordance with the ICC arbitration rules, as they apply at the time the dispute is resolved and not at the time of the signing of the contract, however, under the condition that there is no conflict between these rules and the provisions of the lease agreement and/or Greek law. In the event of such a conflict, the provisions of the lease agreement and of those of Greek law shall prevail. The language of the arbitration proceedings shall be in Greek (Article 23.7) unless the parties otherwise agree. It is explicitly provided that the arbitral award is final and binding, (so a further judicial remedy is not provided) and may be brought before a competent court for recognition and enforcement (Article 23.8). As long as any arbitration procedure lasts, the deadlines set by the parties are suspended, insofar as they relate to each (pending) arbitration procedure. On the contrary, during the arbitration process, the contract itself is not suspended, and the lessee has the right and the obligation to continue with the work provided in the contract (Article 23.9–23.10).

6 Variations of Law 4525/2018 on issues of the dispute settlement resolution mechanism

Law 4525/2018 (regarding the marine area 2 of the Ionian Sea) differs from the other three corresponding laws (4524, 4526, 4527/2018) in terms of how the dispute could be determined by an "Exclusive Expert" before the referral to arbitration and the way the "Exclusive Expert" is appointed. In addition, it provides the possibility of resolving disputes via mediation.[45] The "Exclusive Expert" is hereby appointed by the Parties within fifteen days (Selection Period) from the submission of a written notification by the requesting Party to the "Recipient" party in relation to its intention to refer the dispute for determination to an "Exclusive Expert". The Lessor and the Lessee are bearing equally the cost of the "Exclusive Expert" and those of the process. If the parties fail to reach an agreement on the appointment of the "Exclusive Expert" (during the selection period), it is provided (as an alternative) that he/she shall be appointed within the next fifteen days by the President of an Institute of those included in the definition of the "Exclusive Expert", provided that the President in question is not subject to any conflict of interest. The law provides (in a really broad way) that the "Exclusive Expert" should be a person qualified according to his/her education, experience and professional training in order to determine the matter in question and in addition to be a person generally accepted by the international oil & gas industry, as an expert in the field or areas of specialization related to the dispute. It is also explicitly stated that no one could be appointed as an "Exclusive Expert" if he/she has or may have any interest or duty which is or may be incompatible with his/her role as such. Likewise, no one may be appointed as an "Exclusive Expert" if he/she is or has previously been a director, officer, employee or advisor in relation to any party or any affiliated business of this party. Detailed conditions are therefore laid down in relation to qualifications, potential conflicts of interest and the exclusion of the "Exclusive Expert".

Once the "Exclusive Expert" is appointed, in accordance with the provisions of the law and provided that the parties have mutually agreed in writing "the description" of the dispute and the conditions under which the "Exclusive Expert" shall try to determine it, the Lessor is obliged to immediately notify the "Exclusive Expert" that he/she has been appointed by the parties and to ask him/her to state whether he/she wishes and can accept his/her appointment within five days (Acceptance Period). In the event that the "Exclusive Expert" either refuses to accept or is unable to accept his/her appointment or does not accept it within the "Acceptance Period", then the parties shall choose an alternate "Exclusive Expert" within five days from the termination of the "Acceptance Period". If the Parties fail to agree on the appointment of the "Exclusive Expert"

---

45 See Anna Plevri, "Arbitration in the energy sector (Greek and European institutional framework)", pp. 87–90 (in Greek).

within the required time limit, the matter shall be referred by the parties to the President of the Institute (as described in Article 23.2 (a)) and the procedure shall be repeated until an "Exclusive Expert" who accepts his/her appointment on terms acceptable to all parties is selected (appointed) or agreed. In order for the dispute to be determined by the "Exclusive Expert", each party is required to submit to the other and to the "Exclusive Expert" within 30 days (Submission Period) after the acceptance of the appointment of the "Exclusive Expert", a description of the dispute and any documents supporting and/or justifying its position. The "Exclusive Expert" has a complete discretion to take into account any additional information submitted by any party and/or any other procedural matters not expressly mentioned in the law. The terms of reference under which the "Exclusive Expert" seeks to determine the dispute shall be mutually agreed by the parties, and the criteria under which the "Exclusive Expert" shall reach his/her "decision" should fall strictly within the terms of reference agreed between the parties (Article 23.2 (c)). The "decision" of the "Exclusive Expert" is final and binding on the parties and should be reached within 30 days after the end of the submission period, except in cases of fraud or a manifested error. This "decision", otherwise the conclusion of the "Exclusive Expert", should be written in English and may be eventually referred to arbitration on a point of law and not on a factual matter. The language of the arbitration procedure shall be (according to the law) Greek and English (Article 23.7) unless the parties agree otherwise. This provision is rather ambiguous in the sense that it is not clear whether arbitration proceedings could be conducted in both languages alternatively or parallel and if one of the two languages could prevail. Finally, Article 23.12 provides the possibility for the parties to agree at any time and without prejudice to any other procedures, to refer any dispute to mediation, in accordance with the ICC mediation rules, as those apply at the time of the referral of the dispute to mediation. This way, the ICC mediation rules will be considered as incorporated by referral to Article 23.12 of the contract.

Conclusions

Undoubtedly, the energy sector is an area where conflicts frequently arise at an international level. In order to understand this fact, one should have in mind the economic and energy framework worldwide which contains energy projects that are capital-intensive, a possible economic downfall, increasing lack of energy supplies, plummeting oil prices, rising intra-state energy transit, policies of protecting natural resources and growing environmental concerns. Taking the above into consideration, energy-related companies, states, investors and other stakeholders need to take into account the factors of time and cost in order to choose an appropriate and suiting out-of-courts dispute resolution method such as arbitration, which actually is private justice with many advantages. Naturally, the resolution of any dispute at an early stage is always preferable. There are several reasons which justify the "suitability" of international arbitration to resolve energy disputes. These reasons include the need for a neutral process, the ease of enforcement of arbitral awards, the confidentiality of arbitral proceedings, the speed at which energy disputes can be resolved by arbitration, the cost of arbitration, the choice of forum and panel – meaning the selection of "skilled" arbitrators – the parties' control of the procedure, the flexibility of the procedure and of the rules of evidence, and the availability of interim measures.

To conclude, due to factors such as the complexity of energy disputes where profound technical knowledge is needed, the fact that energy disputes often have a very high monetary value, a strong public interest and a cross-border character, also due to the length and other characteristics of court proceedings and the increasing political significance of energy-industry allocation, ADR methods and especially arbitration is a powerful tool, used by parties internationally in order to protect and safeguard valuable economic interests and rights of the energy sector.

Michalis Mathioulakis*

8 The role of legal disputes and arbitration in EU's energy security: The case of natural gas networks

Abstract: The structural gravitas of energy in all aspects of economic and social life, inevitably deems energy dependency a critical issue, raising concerns of political vulnerability on a national and EU level. This, in turn, often generates a fierce response by the Member States towards practices from energy companies that threaten their energy security. This chapter examines the consequent legal disputes and the utilisation of legal instruments that often lead to arbitration and resulting awards whose size can practically overturn long-designed business plans regarding the flow of natural gas in Europe. Selected legal disputes are examined under the effects they can generate for the energy security in Europe and the countries they affect.

Keywords: Natural gas, energy security, arbitration in the gas networks, take-or-pay cause, Gazprom, PGNiG, Naftogaz, Nord Stream 2, EU Commission

Introduction

The issue of energy security holds a central role in the overall energy policy of the European Union. The concept of energy security includes a broad spectrum of concerns extending from ensuring the infrastructure that supports energy demand, to efficiently responding to abrupt discrepancies in energy supply, and from securing the availability of energy resources to providing affordable prices for energy products. The International Energy Agency, (IEA) defines energy security as the uninterrupted availability of energy sources at an affordable price.[1] For the European Union, where more than 55 % of its energy needs are covered through imports,[2] energy security is heavily related to the control over, and diversification of, energy resources. The structural gravitas

---

*   Michalis Mathioulakis is an Energy Strategy Analyst and the Research Director of the Hellenic Energy Regulation Institute.
1   "Energy Security – Areas of Work", *IEA*, https://www.iea.org/areas-of-work/energy-security (accessed 29 January 2020).
2   "Energy Production and Imports", Eurostat, https://ec.europa.eu/eurostat/statistics-explained/index.php/Energy_production_and_imports (accessed 29 January 2020).

of energy in all aspects of economic and social life, inevitably makes extended energy dependency a critical issue, raising concerns of political vulnerability on a national and EU level. This, in turn, often generates an aggressive response by the Member States towards practices from energy companies that threaten their energy security with the utilisation of legal instruments that often lead to arbitration and consequent awards whose size can practically overturn long-designed business plans. The context of energy security has been broadening throughout the years alongside the reasons that carry the capacity to generate disruptions in the energy supply chain. Political unrest like the Russia-Ukraine natural gas disputes in 2006–2009, instability in the Middle East affecting the global oil supply, natural disasters like the earthquake and tsunami in Japan in 2011, and power system failures like the 2003 blackout in the northeast US, raise overall concerns in Europe over the vulnerability of economic systems and human activity and wellbeing due to energy-related disruptions. The importance of energy security in the EU derives from the Union's growing dependency on energy imports. Despite regressing rates of energy demand, the steep decline in primary energy production in the EU led to an increase of its 'energy dependency rate', the percentage of net energy imports to gross available energy, to more than 55 % in 2017.[3] This, in turn, raises the EU's dependence and vulnerability from third countries, primarily from Russia. According to the latest published official data from Eurostat, Russia accounts for 38.9 % of the EU's imports of coal, 33,7 % of its crude oil imports and 38,7 % of total EU imports of natural gas.[4] Severe dependency from Russian energy sources leads to excess vulnerability of EU's energy systems while simultaneously exposing the Member States to political pressure from the Kremlin. The 2006–2009 Russia-Ukraine conflict and the discrepancy of gas flow towards the EU has been the catalyst to showcase such vulnerabilities.

1 The role of legal disputes and arbitration in natural gas imports for the European energy security

The decline in primary energy production in the EU leads to excessive concerns over the Union's ability to maintain effective policies regarding energy security without raising its dependence from energy imports. To this end, EU's plans

---

3 "Energy, transport and environment statistics", *Eurostat*, https://ec.europa.eu/eurostat/documents/3217494/10165279/KS-DK-19-001-EN-N.pdf/76651a29-b817-eed4-f9f2-92bf692e1ed9 (accessed 29 January 2020).
4 Ibid.

towards the transition to a more sustainable energy mix seem to collide – or at least not align – with the existing and planned infrastructure in natural gas networks. Germany's decision to phase-out its nuclear power generation following the 2011 Fukushima accident lead to the need for accelerated power production from other sources. This, in turn, has helped boost the county's RES power production but simultaneously increased Germany's need for gas-powered production and therefore its dependency and vulnerability from Russian gas imports. As nuclear and coal power production has been a major pillar for energy security in several EU Member States, a phase-out process like the one applied in Germany has not been met in an equally positive manner by several Member States. More importantly, the lengthy dispute resolution process between the Swedish utility Vattenfall against the German government ('Vattenfall AB et al. v Germany') over the latter's decision to completely pull out of nuclear power, creates additional caution in France and Belgium regarding the possibility of phasing out its own – extensive – nuclear power production. Similarly, to the situation with nuclear and coal power production, the decline in primary natural gas production within the EU borders, primarily due to decreasing production in the North Sea and Holland's gas fields, lead to increasing levels of dependency from gas imports. Long-term binding contracts with fixed selling price and volumes, combined with limited routes of gas flows into Europe, have created the conditions for legal disputes and arbitration in the gas sector. In the case of natural gas prices and network routes, the imperative effects of energy security determine the intensity and gravitas of the legal disputes.

2 Disputes and arbitration over gas imports in Europe

Most of the cases of legal disputes in the gas sector in Europe resolving in arbitration are connected to Gazprom, the Russian state-owned oil and gas company. These refer either in disputes regarding the price of gas or the legal framework of gas pipelines running from Russia to transit or end-user countries.

2.1 The OPAL case

The provisions of EU's energy regulation framework have been applied in – both successful and unsuccessful – efforts to deter or reduce the influence of Gazprom, the Russian natural gas giant who is the primary source of Russian gas to Europe. In one of the most recent occasions, a decision in September 2019 by the Court of Justice of the European Union (CJEU) in case T-883/16 "Poland v. Commission" (the "OPAL case") has ruled that Gazprom could not use the extended capacity of the OPAL pipeline, a Gazprom-owned onshore pipeline running across the eastern part of Germany to the Czech Republic as an extension to the, also Gazprom-owned, Nord Stream 1 natural gas pipeline.[5] Since mid-2009, the German Regulator (BNetzA) had granted an exemption to Gazprom from the provisions of Articles 9, 32, 33 and 34 and Article 41(6), (8) and (10) of Directive 2009/73/EC, regarding the OPAL pipeline. The exemption had been granted by the German Regulator for a 22-year period. Under the exemption's conditions, Gazprom could reserve only 50 % of the cross-border capacity of the OPAL pipeline. However, in 2016, in a new decision, the German Regulator allowed Gazprom to use 80 % of the OPAL transmission capacity. Furthermore, since technically Gazprom was the only company able to supply gas to the OPAL pipeline through its entry point, there was no Third-Party demand to book capacity, so the Russian company could reach up to 100 % of the pipeline's capacity.

The BNetzA decision to extend the OPAL capacity should be seen under the framework of Germany's extended need to improve its energy security regarding the flow of natural gas in the country. Germany's policies in phasing-out of coal and nuclear power production, combined with the extended needs of its industrial sector, inevitably lead to an increased need for natural gas. Soon after its 2016 decision, BNetzA requested the Commission to grant an exemption of the OPAL gas pipeline from Third Party Access provisions. In October 2016, the Commission granted the exemption. This led Poland to appeal, arguing that the Commission decision infringes the principles of energy security and energy solidarity. The CJEU in 2019, annulled the Commission decision approving the modification of the exemption regime for the operation of the OPAL gas pipeline, concluding that the Commission's 2016 decision was adopted in breach of the principle of energy solidarity.[6] It is important to notice the Court's reasoning for its decision. The Court argued that "It does not appear

---

5 General Court of the European Union, Press release no. 107/19 of 10 September 2019, "The General Court annuls the Commission decision approving the modification of the exemption regime for the operation of the OPAL gas pipeline", https://curia.europa.eu/jcms/upload/docs/application/pdf/2019-09/cp190107en.pdf.

6 General Court of the European Union, Press release no. 107/19 of 10 September 2019, "The General Court annuls the Commission decision approving the modification of the exemption regime for the operation of the OPAL gas pipeline", https://curia.europa.eu/jcms/upload/docs/application/pdf/2019-09/cp190107en.pdf.

therefore that the Commission has examined what the consequences might be, in the medium term, in particular for Poland's energy policy, of the transfer to the Nord Stream 1/OPAL transit route of a share of the volumes of natural gas previously transported by the 'Yamal' and 'Braterstwo' pipelines, nor that it has weighed those effects against the increased security of supply at the EU level, which it found." The direct link between energy security and the effects it imposes into legal disputes and their settlement in the gas sector is clearly stipulated in this case by the Court's reasoning. The ruling in the OPAL case has been a cornerstone in an issue that has long troubled EU energy policy and has the capacity to generate a multi-layered impact on the further interpretation and application of the rules of EU energy law.

2.2 Gazprom – PGNiG arbitration

An ongoing dispute between Poland's largest gas distributor PGNiG and Gazprom concluded in a decision by the Arbitral Tribunal in Stockholm in favour of the Polish company. The dispute between the two companies was related to the price calculation procedure for gas supplied under the Yamal contract between the companies. The price formula in force resulted in PGNiG paying for gas in prices considerably higher than the market rate. In 2014, PGNiG used the price renegotiation option provided for in the contract, and in May 2015, after the negotiation period expired, submitted the dispute for resolution by the Arbitration Court in Stockholm. In mid-2018, the Arbitration Court issued a partial ruling declaring that the conditions for PGNiG's demand for changes in the price calculation method have been satisfied. In a statement on March 30, 2020, PGNiG, announced that the five-year-long dispute between them and Gazprom ended with a decision by the Arbitral Tribunal in Stockholm that has ruled in favour of PGNiG, concerning the price of gas. Jerzy Kwieciński, the President of the Management Board of PGNiG SA, stated that "The Arbitral Tribunal has sided with PGNiG, thus confirming that the price of gas in the Yamal Contract failed to reflect the price level on the market and was overstated".[7] According to the Tribunal's decision, the price that PGNiG will pay to Gazprom for natural gas should be based on a new price calculation

---

7 "Victory for PGNiG: The Arbitral Tribunal in Stockholm rules to lower the price of the gas sold by Gazprom to PGNiG", *PGNiG News*, 30 March 2020, http://en.pgnig.pl/news/-/news-list/id/victory-for-pgnig-the-arbitral-tribunal-in-stockholm-rules-to-lower-the-price-of-the-gas-sold-by-gazprom-to-pgnig/newsGroupId/1910852.

formula, that better reflects the gas price levels in the EU gas market. The Tribunal's ruling is applicable for a time period beginning in November 2014, where PGNiG sent its contract price review request to Gazprom. Conclusively, Gazprom is required to pay back to PGNiG an estimated USD 1.5 billion, which is the difference between the price calculated based on the new formula and the amounts actually paid by PGNiG from November 2014 to February 2020.

The direct link between the result of arbitration and energy security for Poland can be effortlessly understood in Jerzy Kwieciński comments on the Tribunal's ruling. PGNiG's executive stated that "Diversification of gas supplies remains our constant priority; therefore, we plan to use the funds we will recover from Gazprom to purchase new hydrocarbons deposits". Indeed Poland's volume of liquid natural gas (LNG) imported from Qatar, the US and Norway has been dynamically rising, while in 2018 and 2019, PGNiG signed long-term contracts for the purchase of LNG from the US. The Tribunal's award thus clears the way for a more intensive diversification of Poland's gas imports. Furthermore, Poland is also advancing its plans for the construction of new nuclear power stations to fill the gap created by the gradual closure of it ageing coal fleet. The country expects its first nuclear power station to start operating in 2033. A total of 6–9 GW of nuclear power is expected to be installed by 2043 while according to government plans, Poland expects to have 2–3 GW of nuclear power production by the mid-2030s.[8]

2.3 Gazprom – European Commission notice of dispute

In a similar situation to the OPAL case, Gazprom has been facing legal challenges regarding its Nord Stream 2 pipeline. The pipeline is designed to carry 55 bcm/y of natural gas from the Russian port of Ust-Luga in the Leningrad region, across the Baltic Sea to Germany in the Greifswald area close to the exit point of Nord Stream 1. Following the CJEU's Judgment in the 'OPAL case', Gazprom should have been expected to face difficulties for Nord Stream 2 to comply with the examination of the security of supply criterion stipulated in Article 11 of the 2009 Gas Directive, or the competition and market-liberalisation provisions in Article 36 of the same Directive. The importance of the Nord Stream 2 pipeline for Russia, combined with the serious threats it is perceived to hold over the energy security of several EU Member States, brought Gazprom and the EU in

---

[8] "Poland firms up gas, nuclear plans to fill green power gap", *EURACTIV*, 12 July 2019, https://www.euractiv.com/section/electricity/news/poland-firms-up-gas-nuclear-plans-to-fill-green-power-gap/.

the brink of arbitration in 2019. Following up on the Gas Directive amendment early in 2019 that provided for the rules governing the EU's internal gas market to also apply to pipelines to and from third countries, Gazprom initiated a series of steps against the European Commission itself. In April 2019, Nord Stream 2 CEO Matthias Warnig, sent a "notice of dispute to the EU" under the Energy Charter Treaty, to the, then, European Commission President Jean-Claude Juncker. Gazprom claimed that if the pipeline is not eligible for a derogation from the new rules, "the measure would be discriminatory against Nord Stream 2 as an investor" and the EU may be breaking the ECT.[9]

Escalating its dispute with the European Commission, the Nord Stream 2 pipeline consortium filed a notice in September 2019, asking a tribunal of private arbiters to determine whether the European Union is in breach of its obligations under Articles 10 and 13 of the Energy Charter Treaty.[10] While the amendments in the Gas Directive related to third-party access, tariff regulation, ownership unbundling and transparency aren't directed against any particular project, the Gazprom Company argued that they were designed to stop Nord Stream 2. The Directive extends exemptions for pipelines that are completed before May 23, 2019, which is the date of its entry into force. According to Gazprom, the discriminatory treatment of Nord Stream 2 derives from the fact that it is the only gas import pipeline that cannot benefit from these exemptions since the final investment decision was made before this date, even though significant capital was committed. It needs to be noted, however, that regardless of the company's efforts, until the first quarter of 2020, the pipeline is yet not completed. However interesting the examination of such a case would be, the significant potential risks for either the EU taxpayers or Gazprom, rendered such an arbitration procedure a rather undesired development. The case seemed to be "solved" after the German parliament voted in November 2019 to approve the implementation into national law of the amendments to the EU gas directive with the inclusion in its accompanying report of a special declaration that offers varied interpretations for the 'completed' state of a pipeline, thus making it possible for the German Regulator to deem Nord Stream 2 as being 'completed'.[11]

---

9   "Notification of possible breach of the ECT", *Nord Stream 2*, 12 April 2019, https://www.politico.eu/wp-content/uploads/2019/04/Nord-Stream-2-ECT-letter-legal.pdf.
10 "Nord Stream 2: Pipeline spat with EU evolves into ECT dispute", *International Institute for Sustainable Development*, 17 December 2019, https://www.iisd.org/itn/2019/12/17/nord-stream-2-pipeline-spat-with-eu-evolves-into-ect-dispute/.
11 "German parliament approves implementation of EU gas directive amendments", *S&P Global Platts*, 14 November 2019, https://www.spglobal.com/platts/en/

However, such a decision can still not be taken for granted by the German Regulator (Bundesnetzagentur). A possible denial by Bundesnetzagentur to grand the necessary exemptions to Nord Stream 2, based on its own interpretation of the term "completed", revives the possibility of Gazprom proceeding with arbitration against the European Commission under the Energy Charter Treaty. However, such a decision does not necessarily bring the end of Nord Stream 2, since Gazprom still has the option to simply comply with EU's regulatory framework by setting up an independent transmission operator or system operator or transfer the operation rights of the German section of the pipeline to either one of the existing German transmission system operators. The potential risks from a long and risky arbitration process could, therefore, offer in this case, the proper incentive for a positive resolution of the ongoing dispute between Gazprom and the EU.

2.4 Gazprom – Naftogaz arbitration

Besides the above-mentioned cases of legal disputes and arbitration, Gazprom has also been in the centre of one of the most famous arbitration cases in the gas sector regarding its dispute with Ukraine's national oil and gas company Naftogaz. Initiated in 2014, the arbitration proceedings have become one of the largest commercial arbitration ever. Mutual claims amounted to about USD 125 billion, which threatened to bankrupt both Gazprom and Naftogaz. In a series of decisions, the Arbitration Institute of the Stockholm Chamber of Commerce (SCC) has delivered two awards; the first one on December 22, 2017, concerning a gas supply contract based on the "take or pay" principle, and the second one on February 28, 2018, with respect to a transit contract. The Stockholm arbitration satisfied Naftogaz's claim for compensation of USD 4.63 billion for Gazprom's failure to deliver the agreed volumes of gas for transit. Following the results of the two proceedings, Gazprom was to pay Naftogaz USD 2.56 billion adjusted to a USD 2.1 billion set-off for gas delivered in 2014.[12] On December 22, 2017, the Arbitral Tribunal ruled in favour of Naftogaz on all contentious issues concerning contracts for the supply of gas with Gazprom: pricing, the

---

market-insights/latest-news/natural-gas/111419-german-parliament-approves-implementation-of-eu-gas-directive-amendments.

12 "Historical victory for Ukraine: Stockholm arbitration", *Naftogaz Group*, http://www.naftogaz.com/www/3/nakweben.nsf/0/E62D5C9B21795281C225834B00537D4E?OpenDocument&Expand=2& (accessed at 15 March 2020).

"take or pay" provision, and invalidating other provisions that are ungrounded and contrary to the principles of competition.

The arbitration court dismissed Gazprom's retrospective claims for USD 56 billion in accordance with the "take or pay" provision (which sets up the obligation to pay for unused gas volumes) for 2009–2017. Naftogaz reduced the future mandatory annual volumes of gas purchases from Gazprom from 42–52 bcm to 4–5 bcm, which corresponds to its actual needs for gas imports. Furthermore, Naftogaz revised the contract towards the reduction of gas prices according to market conditions, as well as compensation for overpayment in the periods after April 2014 when Naftogaz initiated a revision of the price. In particular, the price of gas received by Naftogaz in the Q2 2014 is reduced by 27 % – from USD 485 tcm to USD 352 tcm. The arbitrators also indicated that Naftogaz should not pay for the volumes of gas supplied to the temporarily occupied territories of Luhansk and Donetsk regions, as the volume of these supplies could not be determined. In addition, the arbitration has declared invalid some other provisions of the contract, including the prohibition on re-export. So Naftogaz was allowed to resell gas abroad. On February 28, 2018 the Stockholm Arbitration Tribunal ruled in favour of Naftogaz on issues in the dispute with Gazprom regarding the existing gas transit contract. The Tribunal confirmed the violation by Gazprom of its obligations for transit volumes, which according to the contract amount to 110 bcm per year and awarded compensation of USD 4.63 billion. The arbitration award confirmed the legal obligations of Gazprom concerning deliveries under the transit contract. The Tribunal did not support Naftogaz's request for revision of the transit tariff since the review application filed by Naftogaz in 2009 did not meet the procedural requirements. The Tribunal also rejected Naftogaz's demand to review the transit contract in accordance with European and Ukrainian energy and competition legislation, noting that the implementation of regulatory reform in Ukraine is a matter for the Ukrainian authorities and is not within the competence of the Tribunal.

In the next stage of the dispute between the two companies, the Svea Court of Appeal of Sweden rejected in November 2019 an appeal filed by Gazprom against the 2017 decision made by the Arbitration Institute of the Stockholm Chamber of Commerce (SCC). The ruling reinforced the two previous decisions of the SCC, binding the Russian gas company to keep its then gas prices for Ukrainian consumers in line with market standards, cancel its "take-or-pay requirements" and gas re-exports from Ukraine. The Swedish Court found that "the judgement means that the actions of Gazprom are dismissed and that the special arbitration will not be set aside in any way. As part of the assessment, the Court of Appeal found that there was no evidence that there was an error in the procedure".[13] The ruling by the Swedish Court of Appeal came only a few weeks before the existing contract between the two companies was reaching its expiration in December 2019 and amid ongoing negotiations between the two parties for its continuance. This, combined with the fact that the Nord Stream 2 pipeline was not yet completed, raised great concerns for the European Commission since failure to come into a settlement, would pose serious threats to the flow of gas in Europe and the energy security of the EU. Eventually, at the end of December, days before the expiration of the existing contract, a deal was reached at three-way talks between Kiev, Moscow and the European Union. Gazprom agreed to pay USD 2.9 billion to Naftogaz awarded by the Stockholm arbitration in 2018, while Naftogaz agreed to withdraw all its legal claims against Gazprom, including the lawsuit it filed against Gazprom with the SCC, asking to revise transit fees of more than USD 12 billion.[14]

The direct link between the results and critical role of arbitration in the gas sector with EU's energy security is demonstrated in the comments of Maroš Šefčovič, the European Commission's Vice-President regarding the case, in which the Commission played an active role. On the day the deal was announced, Mr. Šefčovič said that "we have secured an uninterrupted transit of Russian gas to Europe via Ukraine for a period of up to fifteen years. Long-term security of supply to European households was my number one priority and we have sent a powerful message that the EU is guaranteeing warm homes, stable markets and competitive prices".[15] Treating the result of the arbitration procedure between two energy companies as a "win" for European energy security stipulates the critical role of arbitration procedures in the energy sector since the awards resulting from arbitration exceed their financial and business aspect and carry the capacity to generate political results.

---

13 "Naftogaz wins lawsuit against Gazprom's appeal in Stockholm arbitration court", *Emerging Europe*, 28 November 2019, https://emerging-europe.com/news/naftogaz-wins-lawsuit-against-gazproms-appeal-in-stockholm-arbitration-court/.
14 "Gazprom, Ukraine Agree on Gas Transit, Settle Legal Disputes", *Bloomberg*, 21 December 2019, https://www.bloomberg.com/news/articles/2019-12-21/gazprom-ukraine-agree-on-transit-volumes-settle-legal-issues.
15 "Everyone is a winner after EU-Russia-Ukraine gas talks", *EURACTIV*, 21 December 2019, https://www.euractiv.com/section/energy/opinion/everyone-is-a-winner-after-eu-russia-ukraine-gas-talks/.

Conclusions

Dispute settlement and arbitration procedures regarding the price and network routes of natural gas in Europe are strongly related to concerns over national and EU energy security. EU's policies towards the transition to a more sustainable energy mix and the protection of the environment from elevated CO_2 emissions, lead to ambitious phase-out policies regarding nuclear and coal power production. This, in turn, raises concerns over the elevated dependency in imports of natural gas from third counties, as well as the stability and affordability of gas prices in Europe. In their effort to increase energy security, the EU Member States and other European countries, as well as the European Commission have been subjects to legal disputes and arbitration procedures, involving primarily the Russian gas giant of Gazprom, regarding the pricing of natural gas and the legal framework of gas pipelines in the European Union. Examining selected cases of legal disputes involving Gazprom and Poland, Ukraine, and the European Commission, this chapter stipulated the direct connection between the struggle for improved energy security, functioning as the underlying driving force in the parties involved, and the utilisation of legal instruments and arbitration whose results and consequent awards have a direct effect in business relations between energy companies in the gas sector but also directly affect the overall energy security status on a national and European level.

Sofia Michelaki[*]

9 Claims of users concerning capacity charges of Natural Gas Transmission System. Case law

Abstract: This chapter provides a presentation of arbitration settlement developments in disputes regarding the binding force of the capacity-booking process in the Greek Natural Gas Transmission System and its resolution regarding the protection of the contracting parties rights in case of an incident of *force majeure*. The question brought before the Arbitral Tribunal due to a dispute between Users in the Greek Natural Gas Transmission System and their Supplier, concerned their obligation to pay capacity charges in the event of an incident of *force majeure* preventing the use of gas quantities. According to the legal framework that existed during the time of the dispute, TSO's transmission services comprised of two distinctive charges, i.e. the capacity charge which is a fixed charge, and the commodity charge, which is a variable one, depending on the transported gas quantities. The Arbitral Tribunal resolved that an incident of *force majeure* does not affect the firm capacity that has already been booked, so the respective capacity charges are still due, regardless of the *force majeure* incident actually occurring, or not. The *force majeure* incident can only affect the actual use of said capacity, and only the respective commodity charges are to be considered as inactive. The *force majeure* incident can only affect the use of said capacity, and only the respective commodity charges can be released.

Keywords: Capacity charges, commodity charges, booked firm capacity, derivative obligations of users

Introduction

This chapter examines the arbitration settlement of disputes concerning the binding force of the process of booking of firm capacity in the Greek Natural Gas Transmission System and its resolution regarding the protection of the contracting parties rights in case of an incident of *force majeure*. According to the Arbitral Tribunal, the process of booking firm capacity is considered to be similar to that of an insurance policy contract, whereas the agreed fee for the service is due from the client's side, regardless of the occurrence of an event

---

[*] Sofia Michelaki, General Legal Counsel of the "Hellenic Gas Transmission System Operator (DESFA S.A.)".

that triggers the use of said service, since there is a commitment of the provider of the service that the said service is available to the client, upon request from his side. In such a context, in the event of an incident of *force majeure* occurring, the respective capacity charges are still due, regardless of the force majeure event actually occurring or not, with only the commodity charges to be contested, if no actual transfer of gas volumes took place due to this *force majeure* incident.

1 Recent arbitration cases regarding capacity charges

Regarding natural gas transmission, two interesting cases were brought before the Arbitral Tribunal from users of Natural Gas Transmission System, concerning capacity charges. In both cases, claimants had a contract with a natural gas supplier for a bundled product, i.e. the purchase of Natural Gas for their units and the transport of the respective volumes to their units via the existing Natural Gas Transmission System. Both claimants entered into Natural Gas supply contracts with a supplier of gas (supply contract), while the Supplier entered into a transmission agreement with the TSO for booking the necessary capacity into the Transmission System (Standard Transmission Agreement) to fulfil the supply contract. It needs to be mentioned that the booking of capacity is defined as the right to use the Transmission System (network) for a specific period and specific quantities expressed, then, in MWh, now in KWh. During their contract, the claimants invoked incidents of *force majeure* in their industrial units and requested their Supplier to release them (among others) from their obligations to pay the transmission charges for the duration of the *force majeure* incident. It is to be noted that TSO's transmission services, comprised, at the time of the dispute, of two distinctive charges; the capacity charge which is a fixed charge, and the commodity charge which is variable depending on the transported gas quantities. The Supplier that had already paid the transmission charges to the TSO disputed the *force majeure* incident in both cases and each claim was brought before an Arbitral Tribunal in Greece.

Concerning the transmission charges, the Supplier argued that the *force majeure* event does not affect already booked firm capacity but can only affect the use of said capacity. Thus, in case of a justified *force majeure* event, the network user can be released only from commodity charges that concern the transfer of gas quantities. The respective capacity charges are still due, regardless of whether the *force majeure* event actually occurred. The above legal arguments were accepted in both cases by the Arbitral Tribunal, regarding capacity charges. Specifically: The Arbitral Tribunal concluded in both cases that the booking of firm capacity should be considered as similar to an insurance policy contract where the fee is due by the client independent of the occurrence of an event that triggers the use of the service. It is an agreement that provides the commitment of the insurance provider that the service is available at the time it will be requested. So, according to the case law, even in case the *force majeure* event occurs, capacity charges are due if firm capacity is booked, independently on the actual transfer of the gas volumes (commodity charge). If no actual transfer of gas volumes occurs, a release is possible only for commodity charges.

2 Legal analysis

2.1 Legislative Framework

DESFA's activities, as applies to Natural Gas Operators in general, are regulated ex ante, in detail, based on transparent rules known in advance, while invoices are approved by the Regulatory Authority and the Model Framework Agreements for the Transmission of Natural Gas (Transmission Agreement) and the Use of DESFA's LNG Facility on the Revithoussa Island. DESFA's activities are, as arises from the institutional framework of EU and domestic law, fully and comprehensively regulated on an ex lege and ex ante basis, and, as such, DESFA acts in alignment with the law. According to the domestic regulatory framework, DESFA's role as an Operator, provides all interested parties that comply with statutory requirements access to the National Natural Gas System (NNGS) and, to this end, DESFA is responsible for the operation, exploitation and management of high-pressure pipelines and related facilities that make up the afore-mentioned system. DESFA is not, and cannot be, related in any way whatsoever with the natural gas supply and trade market where its customers/Users of the National Natural Gas Transmission System (NNGTS) and of the LNG Facility are active. In order to gain NNGTS access, Transmission Users must conclude with DESFA a template framework agreement for transmission on pre-defined terms, which is pre-approved by the bodies responsible according to the law (currently, such agreement is approved by virtue of a decision by RAE), whereas neither DESFA nor the Transmission User (counter-party) may deviate from the terms of the model agreements, given that the framework thereof is expressly determined by a regulatory act. According to the Transmission Agreement, DESFA is obliged to provide transmission services, namely, take delivery of Natural Gas from the Transmission User, from one or more points of the Transmission System, transfer said quantity of Natural Gas through the system, deliver the natural gas quantities to one or more points, pursuant to the specific agreement terms, and perform the necessary measurements through the metering stations at entry and exit points. To this end, the Transmission Agreement shall determine, inter alia, the capacity (in the domestic Greek regulatory framework it is referred to as Transmission Capacity) which DESFA undertakes, based on contract, to reserve in the Transmission System, upon request, and on behalf of, the Transmission User. As defined in the legislation in force, firm capacity means the natural gas transmission capacity which the Operator guarantees, based on contract, that it is not intermittent; firm services means the services provided by the Operator in relation to the firm capacity and contracted capacity means the capacity made available by the Operator to the User, based on a transmission agreement. Specifically, with regard to Capacity Reservation, this shall take place on the terms and conditions stipulated by law and the Transmission Agreement. Further, according to the Transmission Agreement, the Transmission User shall be obliged to pay a monthly fee to the Operator, in exchange for the provision of the contracted services by the latter. This fee shall be determined and invoiced as pre-defined, according to the provisions of the model agreement and the approved invoices, with DESFA not being entitled to deviate from the said invoicing method. It is pointed out that during the contested period, transmission services were invoiced as the sum of capacity reservation charges, which were fixed and transmitted quantity charges, which were variable. Firm capacity reservation represents a guarantee provided by the Operator that same can indeed make the specific capacity available and that it keeps its facilities in a state of necessary readiness to this end. According to the afore-stated provisions, this contractual obligation shall not be provided for free and is to be compensated through the capacity reservation charges.

2.2 Factual background with regard to the case

Around the late 2000s (2000–2010), a natural gas supply and trading company had concluded with the Natural Gas Transmission System Operator a transmission agreement based on which, the Supplier reserved capacity in the Transmission System on behalf of two of its customers in order to supply them with natural gas. At the time in question, an extended mechanical failure occurred at the facilities of said customers, which lasted several months; as a result, said customers were not in a position to get supplied with natural gas quantities for their facilities. At the time in question, these customers, who had notified their natural gas supplier of their inability to take delivery of natural gas due to technical issues, disagreed with the charges imposed by the Supplier and the dispute between said Supplier and each one of said customers gave rise to arbitration proceedings, by means of which the customers contested, among others, the capacity charges for the duration of the failure which had been imposed by their natural gas supplier/counter-party within the scope of the supply agreement concluded between them. Arbitration Courts based their awards on the entire web of contractual relationships (2011–2012 Awards) between the parties. However, on the vital issue of firm capacity charges, the Courts ruled on the obligation to pay the firm capacity charges irrespective of the transmission and delivery of natural gas, as follows: (a) that the price at which the natural gas was sold to the customer/counter-party of the supply and trade company, namely, the natural gas transmission charges (Transmission Charges), includes the capacity charges (Firm Capacity Charges) and the charges for the transmitted quantity of Natural Gas (Commodity Charges), pursuant to law provisions, (b) that Firm Capacity Charges are different to and – as it stands to reason – they precede Commodity Charges, (c) that only the latter requires delivery and are calculated based on delivered quantities, not the former. With regard to the statutory and contractual framework of the Firm Capacity Charges, the Courts concluded that irrespective of the transmission and delivery of natural gas, capacity charges are a separate service and are "in any case" payable. Subsequently, the natural gas supply and trade company notified its counter-party, the Transmission System Operator, of the dispute that had arisen with its customers and of the respective arbitration awards that had been issued, and raised a claim concerning the capacity charges that had been imposed and paid by the company within the scope of the contractual relationship between the same and the Transmission System Operator for the period for which it was held by the Courts that the company had no right to impose charges on the customers. The natural gas supply and trade company claimed that the charges paid for the capacity reservation in the natural gas transmission system should be reimbursed to the same, since the said reservation was not made use of, given that due to failure of its customers' facilities, no natural gas quantities were transmitted during the period in question, specifically: (a) that the consideration paid for the capacity charges of the period in question, pertains to a service that was not provided by the Transmission System Operator to the company, since, in the period in question, no natural gas quantities were transmitted, due to failure at its customers' facilities (third parties who had a contract with the company), and (b) that, although transmission services in the period in question were paid as a sum of the firm capacity charges and the commodity charges, however, they represented a single service, namely natural gas transmission, and that, insofar as the capacity charges do not pertain to a stand-alone service but are linked to natural gas transmission in this case, the counter-performance (natural gas transmission) did not take place and, as such, the consideration is not payable. For these reasons, the company claimed that the amounts paid by the same must be reimbursed to it as an alleged "undue consideration", otherwise as "contractual compensation" or, alternatively, based on the provisions on "unjustified enrichment".

On its part, the Transmission System Operator claimed that: (a) In the dispute period in question, as also applies today, the provision of natural gas transmission services for the delivery of natural Gas to Transmission System Users took place, and still does, based on template agreements pre-approved by RAE and based on a single invoice. (b) The approved invoices for the use of the Transmission System were based on a fundamental principle concerning the recovery of required revenue for the Transmission System Operator and the generation of a reasonable profit, and the required revenue approved in this regulated manner aims, on the one hand, to recover the investment costs of the operators, and on the other, to recover the infrastructure operating costs, so that infrastructures can operate smoothly without failures. Specifically, this applies in order for infrastructure costs incurred for the operation of pipelines, metering stations, compressors, and other facilities to be written-off gradually and for their operating costs to be recovered. Furthermore, the aforementioned required revenue aims to a reasonable profit, which, if exceeding the approved limit, is returned, as far as the surplus amount is concerned, to the market. (c) The template transmission agreement determines, inter alia, the firm capacity (in this case) reserved by the Transmission System Operator on behalf of its counter-party/Transmission User, based on which the Operator guarantees the maximum natural gas quantity which it can receive from a User at an entry point or deliver to the User at an exit point of the Transmission System, during a Natural Gas day. (d) The firm capacity reservation charges pertain to charges that are not linked to the commodity (the natural gas quantity transmitted), since, as elaborated above, regarding capacity reservation, is the Operator's obligation to make a specific capacity available and keep its facilities in a state of necessary readiness throughout the effective term of the agreement, excluding facility maintenance periods. Such service is not provided free of charge but is compensated through capacity charges. (e) Capacity reservation precedes commodity transmission; it is performed upon the agreement being signed, and it is a prerequisite for the transmission to take place. This is a charge precisely similar to the fixed charge imposed on consumers being supplied with electric power or natural gas. A requirement for consumers to be supplied with electric power or natural gas is their connection with the grid and assurances as to the integrity of the connection. The required capacity is ensured throughout the effective term of the supply agreement. The same applies to the capacity reservation. With the Transmission Agreement, a User primarily reserves capacity, meaning they reserve the System's capability to transmit the maximum quantity they wish to receive and, subsequently, natural gas transmission takes place. It goes without saying that if the Transmission System has no further capacity, Users may not reserve capacity and transmit natural gas and, in that case, congestion management procedures are applied. As a consequence, capacity charges are not incidental but rather a basic, stand-alone and separate charge serving as a consideration in order for the Transmission User to be able to maintain its capability for natural gas transmission on a daily basis, whenever the User wishes, with the maximum limit corresponding to reserved capacity. (f) As such, according to the Transmission Agreement, the User shall be obliged to pay a monthly fee to the Operator, in exchange for the provision of the contracted services by the latter. (g) By contrast, commodity charges are charges pertaining to the quantity of the natural gas that is actually transmitted to customers. Capacity charges may be accurately compared to the fixed charges comprised in the electricity bill whereas commodity charges may be compared to electricity consumption charges for the period covered by the respective bill.

Conclusions

Recent resolutions on arbitration settlement, have enormously strengthened the existing legal protection framework for international investors in the European energy sector, with relevance to an ambiguous and crucial legal matter. According to said resolutions, the firm capacity charges – already booked and agreed between the contracting parties for the capacity reserved by the Operator on the User's behalf, based on the relevant provisions of the existing Transmission Agreement – remain, by any means, unaffected in an incident of *force majeure,* in a manner so that the respective capacity charges are still due, regardless of the *force majeure* incident actually occurring, or not. Thus, the aforementioned Arbitral Tribunal Resolution succeeded in bringing a sense of equilibrium and stability in the Energy market, by irrevocably solving a legal dispute, in such a manner that allows the investor much needed "breathing space", with regards to the profitability of his investment, which, after all, is always the most decisive factor in the area of investing.

Peter Mussaeus & Adriane Mönch[*]

10 Germany's electricity price zones and their delineations

Abstract: This chapter begins with a short overview of current energy topics in Europe. First, the General Court's decision on capacity bookings is mentioned, followed by the phasing out of coal-fired power plants and the plans to reform the Energy Charter Treaty. The authors then focus on electricity price zones. As a first step, the system of bidding areas is explained. The common market with a single wholesale price for the same electricity product is presented using the example of Germany. It is elucidated how the difference in the electricity production in North Germany and the consumption in South Germany not only leads to a bottleneck in Germany itself, but also to border-crossing loop flows in Poland and the Czech Republic. Against this background, the Article 14 Regulation (EU) 2019/943 is introduced and examined, as it specifies new procedures regarding the delineation of bidding zones. It further establishes the competence of the European Commission to make an ultima ratio decision, causing a revival in the discussion about the German separation into two different bidding areas. In the next step, the directive is set in connection with § 3a StromNZV, and legal consequences are pointed out. Especially a violation of Article 194 TFEU and Article 2 (2) TFEU, leading to a blocking effect on § 3a StromNZV is taken into consideration. The chapter ends with a conclusion and the recommendation to further extend the electricity network.

Keywords: Electricity price zones, bidding areas, loop-flows, Article 14 Regulation (EU) 2019/943, capacity booking, phasing out of coal-fired power plants, Energy Charta Treaty

Introduction

Due to constantly progressing innovations in technology, economy and regulations, energy law always offers new challenges which must also be implemented on a practical level. Energy policy has become increasingly

---

[*] The author is lawyer and partner of PwC Legal AG Rechtsanwaltsgesellschaft and specialized in European and German energy law. Adriane Mönch is a scientific assistant at PwC Legal AG Rechtsanwaltsgesellschaft with a focus on Energy law. The article is based on the presentation of Peter Mussaeus for the 4th Athens Conference on European Energy Law.

European in recent years as well.[1] In the following article, current topics of European energy law will be examined and explained in more detail with a focus on electricity price zones and the new Regulation (EU) 2019/943.

1 The General Court's decision on capacity bookings

The General Court annulled the Commission's decision,[2] approving the modification of the exemption regime for the operation of the OPAL gas pipeline.[3] The German federal network agency "Bundesnetzagentur" (BNetzA) had allowed a greater usage of the OPAL natural gas pipeline in Germany. OPAL is the extension of the Nord Stream pipeline. Initially, Gazprom was allowed to use 50 % of the OPAL pipeline for reasons of security of supply. With the decision of the European Commission, this capacity permission was extended. Poland and the interveners Latvia and Lithuania applied for an annulment. They argued that the Commission's decision infringed inter alia the principles of energy security and energy solidarity[4] because it enabled Gazprom and undertakings in the Gazprom group to redirect additional volumes of gas into the EU market by fully exploiting the capacities of the Nord Stream 1 pipeline. The reduction in gas transport via the Yamal and Braterstwo gas pipelines could lead to less supply security in Poland and significantly hamper the diversification of supply sources. The General Court found that the principle of solidarity is not referred to in the 2016 decision and that it does not appear that the Commission, as a matter of fact, conducted an examination of that principle. It was the Commission's responsibility to verify in the original decision that the planned investment met this criterion.

---

1 Matthias Schmidt-Preuß, in: *Berliner Kommentar zum Energierecht*, ed. Franz Jürgen Säcker (Frankfurt am Main: Fachmedien Recht und Wirtschaft, 2018), Vol. 1, Part B rec. 1.
2 European Commission, Decision on review of the exemption of the Ostseepipeline-Anbindungsleitung from the requirements on third party access and tariff regulation granted under Directive 2003/55/EC, of 28 October 2016, C(2016) 6950.
3 General Court, judgment of 10 September 2019, case T-883/16, ECLI:EU:T:2019:567, *Poland v. Commission*.
4 Council Regulation 73/2009, OJ L 30, 19 January 2009, 16, Article 36 (1) lit. a in conjunction with Treaty on the Functioning of the European Union, OJ C 326, 26 October 2012, Article 194 (1).

2 Phasing out of coal-fired power plants

Another long-term discussion is the phasing out of coal-fired power plants, which countries such as Germany, France, Belgium, Austria, Netherlands, Scandinavia and Portugal are striving for. In Germany, the Commission of "Growth, Structural Change and Employment" recommended the end of coal-fired power plants in Germany by 2038 the latest whilst showing how structural change can succeed in the affected regions.[5] While the German government is planning a coal phase-out law,[6] the European State Aid has no specific regulations on closure aid for coal-fired power plants.[7] The Council's decision on state aid to facilitate the closure of coal mines only refers to non-competitive ones.[8] Therefore, the regular state aid rules are valid concerning subsidy programs on decommissioning coal power plants. One of the main issues will be the mitigation of the effects promoting the affected areas.[9] Consequently, the Commission's Regulation (EU) No. 651/2014 of June 17, 2014, declaring certain categories of aid compatible with the internal market in application of Articles 107 and 108 of the Treaty might apply in some cases as well. The regulation exempts groups of state aids such as – inter alia – urban development from the notification requirement. In Germany, requirements for an exit from coal would have to be based on the requirements of the content and barrier provisions of Article 14 GG. Because in this case, there would be no expropriation due to the lack of a goods procurement process according to the case-law of the German Federal Constitutional Court (Bundesverfassungsgericht, BVerfG),[10] on the nuclear

---

5 Kommission "Wachstum, Strukturwandel und Beschäftigung", "Abschlussbericht", 31 January 2019, https://www.bmwi.de/Redaktion/DE/Downloads/A/abschlussbericht-kommission-wachstum-strukturwandel-und-beschaeftigung.pdf?__blob=publicationFile&v=4 (accessed 14 November 2019).
6 German government on its website, "Was tut die Bundesregierung für den Klimaschutz?", https://www.bundesregierung.de/breg-de/themen/klimaschutz/bundesregierung-klimapolitik-1637146 (accessed 18 December 2019).
7 Clemens Holtmann and Patricia S. Stöbener de Mora, "Der beihilferechtliche Werkzeugkasten für den Umgang mit der 'Kohle' für den Kohleausstieg", *Europäische Zeitschrift für Wirtschaftsrecht* (2019), 485, 486.
8 Decision 2010/787/EU of the Council of 10 September 2010 on State aid to facilitate the closure of uncompetitive coal mines.
9 Walter Frenz, "Kohleausstieg und Beihilfenverbot", *EnergieRecht* (2019), 193.
10 German Federal Constitutional Court, judgment of 6 December 2016, case 1 BvR 2821/11, *Neue Juristische Wochenschrift* (2017), 217.

phase-out.[11] The mere authorization to construct and operate a power plant cannot be a position of ownership protected under Article 14 GG, although it may create confidence in a certain legal position.[12] The challenge of phasing out coal is to reconcile climate protection and carbon dioxide reduction on the one hand with structural measures and security of supply on the other.

3 The Energy Charter Treaty

The EU plans to reform and modernize the Energy Charter Treaty with special regards to investment protection, climate and disputes between investors and states.[13] This is an important step towards more reliability of the energy market. The Energy Charter Treaty came into force in 1998, and most of the regulations have not been changed since. This is, according to the European Commission, problematic, because the rules on investor protection lack modern standards and are no longer adequate.[14] The member states discuss to which extend arbitration proceedings under the Energy Charter could also be affected by the ECJ ruling in the *Achmea* case,[15] and be inadmissible in the future.[16] The judgment focused on the question of whether arbitration clauses in intra-Union bilateral investment treaties affect the uniform interpretation and application of Union law. Arbitral tribunals also decide on Union law. However, they cannot refer to the ECJ for a preliminary ruling because they are not in a court system of a Member State. In the absence of such control by national courts, the ECJ concluded that Articles 344 and 267 TFEU should be interpreted as precluding

---

11 Cornelia Ziehm, "Das Urteil des Bundesverfassungsgerichts zum Atomausstieg: Konsequenzen für den Kohleausstieg", *Zeitschrift für Neues Energierecht* (2017), 7 et seq.
12 German Federal Constitutional Court, judgment of 6 December 2016, case 1 BvR 2821/11, *Neue Juristische Wochenschrift* (2017), 217, rec. 231 et seq.
13 European Commission, Recommendation for a Council decision authorising the entering into negotiations on the modernisation of the Energy Charter Treaty, COM(2019) 231 final of 14 May 2019.
14 Ibid., p. 1.
15 European Court of Justice, judgment of 6 March 2014, case C-284/16, ECLI:EU:C:2018:158, *Achmea*; see also the comment of Boris Scholtka, "Investitionsschutzrecht: Schiedsklausel in unionsinternem Investitionsschutzabkommen", *Europäische Zeitschrift für Wirtschaftsrecht* (2018), 239 et seq.
16 Patricia S. Stöbener de Mora, "Energierecht: Kommissionsvorschläge zur Modernisierung des Energiechartavertrags", *Europäische Zeitschrift für Wirtschaftsrecht* (2019), 483.

arbitration clauses in intra-EU bilateral investment treaties. The Energy Charter Treaty contains a clause concerning arbitration processes in Articles 26 and 27. As the national law and therefore European law may also be incorporated according to Article 26 of the Energy Charter Treaty, a reform to clarify the relation between the ECJ and arbitration processes seems necessary.[17] Currently, the Commission considers that this Treaty cannot be used as a basis for settling disputes between EU investors and EU Member States.[18]

4 Electricity price zones

A vital topic in the EU is the distribution and management of so-called electricity price zones (also: bidding zones). As a result of the 'Winter Package',[19] the discussion about electricity price zones revived and is now – as new rules were introduced – more important than ever on a European level. The topic is presented using the example of Germany as it shows the legal, political and economic dimensions that are typical for European Energy topics. Since June 5, 2019, the new Regulation (EU) 2019/943[20] has been in force. Among other topics, it contains new regulations for the composition of electricity price zones. In order to explain its significance, the nature and the development process of these zones will be discussed first. An electricity price zone (also: "bidding zone") is a common market with a single wholesale price for the same electricity product.[21] Electricity price zones are spread across Europe. National borders are not a precondition for the delineation of these bidding zones, although some are defined this way (e.g. France).[22] Some countries even apply several

---

17 See also on the consequences of the *Achmea* decision: Patricia Nacimiento and Sascha Bauer, "Das *Achmea*-Urteil des EuGH – und nun?", *Betriebs-Berater* (2018), 1347, 1348.
18 European Commission, Communication from the Commission to the European Parliament and the Council – Protection of Intra-EU Investment, COM(2018) 547 final of 19 July 2018.
19 European Commission, Communication from the Commission to the European Parliament and the Council – Clean Energy for All Europeans, COM(2016) 860 final of 30 November 2016.
20 Regulation (EU) 2019/943 of the European Parliament and of the Council of 5 June 2019 on the internal market for electricity, OJ L 158, 14 June 2019, 54.
21 Sabrina Kuhlemann, in: *Berliner Kommentar zum Energierecht*, ed. Franz Jürgen Säcker, Vol. 3, Article 32 Regulation (EU) 2015/1222, rec. 1.
22 Alexander Lüdtke-Handjery, *Energierecht Kommentar*, eds. W. Danner and C. Theobald (München: C.H. Beck Verlag, 2019), Chapter 15 StromNZV rec. 8.

bidding zones, each with different electricity prices.[23] Why are electricity price zones useful? In Germany, for example, the wholesale price is the same in the North and the South.[24] Figuratively speaking, it is the illusion of a copper plate that transports electricity without any losses.[25] However, the reality is often different: Due to bottlenecks, feed-in management, re-dispatch measures and the delayed expansion of the network, electricity costs can vary from region to region – this is also made clear by the different prices charged by the electricity providers. The end consumer pays the additional costs as a part of the network fee.[26] In order to demonstrate the differences and to match the market design, electricity price zones are created to acknowledge the different network conditions. The advantage is a reduction of bottleneck management and network extension.[27] Instead of executing re-dispatch measures, countries with several bidding zones can adapt the electricity price for the overloaded area while setting economic incentives for the regulation of feed-in and feed-out.[28] Electricity price zones that consider the distribution of supply and demand are a cornerstone of the market-based energy trading and prerequisite to fully exploit the potential of capacity assignment methods, including a flow-based approach.[29]

4.1 The problem of loop-flows

A lot of renewable energy is produced in North Germany, but it needs to be transported to South Germany where there is a lot of consumption but little

---

23 Ibid., rec. 7 et seq.
24 German Federal Government, "Einheitliche Strompreise sichern", 27 December 2017, https://www.bundesregierung.de/breg-de/aktuelles/einheitliche-strompreise-sichern-451908 (accessed 14 November 2019).
25 Carsten König and Max Baumgart, "Der EU-Binnenmarkt und die einheitliche Stromgebotszone in Deutschland", *Europäische Zeitschrift für Wirtschaftsrecht* (2018), 491.
26 German Federal Network Agency, resolution of 27 November 2009, BK8-09-003 to BK8-09-006.
27 See the expert report recommending the split of bidding areas for structural and permanent bottlenecks Frontier Economics/Consentec, "Bedeutung von etablierten nationalen Gebotszonen für die Integration des europäischen Strommarkts" (London: Frontier Economics, October 2011), p. 49.
28 Jan von Aengenvort and Hendrik Sämisch, "Die Illusion einer Kupferplatte #2", *Next Kraftwerke (blog)*, 19 September 2014, https://www.next-kraftwerke.de/energie-blog/preiszonen-modell (accessed 12 November 2019).
29 Regulation 2019/943.

production. Germany's transportation networks do not have enough capacity for the transport yet, because the German network extension is progressing far too slowly.[30] Under this aspect, the phenomenon of so-called loop-flows has occurred in the past, especially with the neighbouring states of Poland and the Czech Republic.[31] Starting in North Germany, the electricity was transported automatically due to the bottlenecks in Germany to Poland and/or the Czech Republic, because electricity always takes the path of the least resistance. In those countries, the network was used and strained without any profit for the trade. From Poland or the Czech Republic, the electricity then got back into the German network. This caused Poland and the Czech Republic to install phase shifters at the border, stabilizing their own networks and blocking border-crossing electricity-flows.[32] The example of Germany and its neighbour states is applicable to other EU-member states as well, e.g. Denmark, Norway and Sweden have already complained to the European Commission about the German network.[33] Austria and Germany had been in the same bidding area for more than 15 years before a separation took place on October 1, 2018.[34] Consistent bottlenecks, re-dispatch measures and loop-flows caused the ACER (Agency for the Cooperation of Energy Regulators) in 2016 to mandate the split of the common bidding area.[35] While the separation of the electricity zones was

---

30 "Monitoringbericht 2018", German Federal Network Agency/German Federal Cartel Office, 28 November 2018, fig. 42 p. 132.
31 Peter Mussaeus, Christoph Sänger and Paula Friederike Linden, "Die Neuordnung der Strompreiszonen – ein ungeliebtes Kind am Horizont", *Recht der Energiewirtschaft* (2017), 172 et seq.; Carsten König, "Die Pflicht zur Umsetzung eines Market Splittings in Deutschland", *Zeitschrift für das gesamte Recht der Energiewirtschaft* (2013), 451, 452.
32 German Federal Government, "Deutsch-polnische Stromflüsse werden geregelt", 17 March 2014, https://www.bundesregierung.de/breg-de/aktuelles/deutsch-polnische-stromfluesse-werden-geregelt-388716 (accessed 14 November 2019).
33 Markus Kahles and Fabian Pause, "Die finalen Rechtsakte des EU-Winterpakets 'Saubere Energie für alle Europäer'", *EnergieRecht* (2019), 47, 52.
34 German Federal Network Agency/E-control, "Trennung der deutsch-österreichischen Strompreiszone", 1 October 2018, https://www.bundesnetzagentur.de/SharedDocs/Pressemitteilungen/DE/2018/20181001_E-Control.html (accessed 14 November 2019).
35 Agency for the Cooperation of Energy Regulators, "Decision of the agency for the cooperation of energy regulators no. 06/2016", 17 November 2016, https://www.acer.europa.eu/Official_documents/Acts_of_the_Agency/Individual%20decisions/ACER%20Decision%2006-2016%20on%20CCR.pdf (accessed 4 June 2020).

well prepared, and the effects small, except for a slight increase in prices,[36] the decision to separate in itself provided sufficient grounds for legal proceedings.[37] Austrian companies criticize the decision, arguing that the German transmission system operator TenneT is now abusing its dominant market position.[38] By introducing bottleneck management at the Austrian-German border, the market is disturbed, as the bottlenecks are actually within Germany and not at the border. The market is therefore distorted in an anti-competitive manner. Several companies have submitted a joint application for the elimination of abuse of a dominant market position to the Vienna Higher Regional Court.[39] In the meantime, the European General Court has declared the order from ACER to separate the Austrian-German electricity price zone null and void.[40] In summary, the Court stated that as soon as an amendment to the terms is requested within the 6-month period under Article 9 (10) Regulation (EU) 2015/1222, the 2-month period under Article 9 (12) Regulation (EU) 2015/1222 applies, while ACER has no competence to decide. At that time, ACER lacked the authorization. The plaintiffs had argued, among other things, that ACER had not waited for the 2-month period: In the present case, E-Control had submitted an amendment since May 13, 2016, to the original joint proposal of the transmission system operators to determine the capacity calculation, which

---

36 German Federal Network Agency/E-control, "Trennung der deutsch-österreichischen Strompreiszone", 1 October 2018, https://www.bundesnetzagentur.de/SharedDocs/Pressemitteilungen/DE/2018/20181001_E-Control.html (accessed 12 November 2019).

37 See for example General Court, judgment of 24 October 2019, case T-332/17, ECLI:EU:T:2019:761, *Energie-Control Austria für die Regulierung der Elektrizitäts- und Erdgaswirtschaft (E-Control) v. ACER* and General Court, order of 5 June 2018, case T-333/17, ECLI:EU:T:218:350, *Austrian Power Grid AG and Vorarlberger Übertragungsnetz GmbH v ACER*.

38 "Klage gegen Aus für Strompreiszone", *Wiener Zeitung*, last modified 8 January 2019, https://www.wienerzeitung.at/nachrichten/wirtschaft/oesterreich/1011244_Klage-gegen-Aus-fuer-Strompreiszone-mit-Deutschland.html (accessed 14 November 2019).

39 "Gemeinsam gegen die Trennung der deutsch-österreichischen Strompreiszone", *APA-OTS*, Press release of 8 January 2019, https://www.ots.at/presseaussendung/OTS_20190108_OTS0122/gemeinsam-gegen-die-trennung-der-deutsch-oesterreichischen-strompreiszone (accessed 12 November 2019).

40 General Court, judgment of 24 October 2019, case T-332/17, ECLI:EU:T:2019:761, *Energie-Control Austria für die Regulierung der Elektrizitäts- und Erdgaswirtschaft (E-Control) v. ACER*.

aimed to abolish the border from the German-Austrian electricity price zone. The 6-month-deadline of Article 9 (10) Regulation (EU) 2015/1222 would have expired on May 17, 2016.

If such an amendment request is submitted, the competent national authorities must decide on the amended common proposal within two months following its submission. If the regulatory authorities cannot agree within this 2-month-period, the ACER is allowed to issue a decision on the *amended terms* within six months. In this particular case, ACER made a decision based on the *original* joint proposal of the transmission system operators although an amendment request had been submitted. The EGC confirmed,[41] that according to Article 9 (12) of Regulation (EU) 2015/1222, as soon as an amendment application is submitted within the deadline, ACER does not acquire the competence to decide under Article 9 (11) of Regulation (EU) 2015/1222. There had been no transfer of the task to ACER either. For this reason, ACER had not been empowered to decide on the determination of the electricity price zones since the submission of the amendment. It remains to be seen whether the decision will become final. Against this background, it seems at least possible that Germany could be divided into two bidding areas in the future. Such discussions have been taking place since the ACER-decision.[42] The bottleneck from North to South has intensified after the installation of phase shifters. While trade relations may be simplified, the consequence would be low electricity prices in North Germany and high electricity prices in South Germany. So, the German energy market finds itself in a kind of standoff, a no-win situation for the government, energy companies and – most important – consumers. In such situations, often unexpected regulations step in.

4.2 The new Article 14 Regulation (EU) 2019/943

The new Article 14 Regulation (EU) 2019/943 is to be inserted in this initial situation. It forbids structural bottlenecks in a bidding area. The electricity price zones shall not have structural bottlenecks unless they do not affect neighbouring bidding zones, or, as a temporary exception, their effects on neighbouring bidding zones shall be mitigated by relief measures and such structural bottlenecks shall not result in a reduction of cross-zone trading

---

41 Ibid., rec. 45.
42 Peter Mussaeus, Christoph Sänger and Paula Friederike Linden, "Die Neuordnung der Strompreiszonen – ein ungeliebtes Kind am Horizont", *Recht der Energiewirtschaft* (2017), 172, 173 et seq.

capacity in accordance with the requirements of Article 16 Regulation (EU) 2019/943. It further states that their borders must be oriented by long-term bottlenecks. The bidding areas must offer the highest possible economic efficiency and many options for cross-zone trade while ensuring the security of supply. Every three years, the ENTSO-E draws up a report verifying any existence of bottlenecks, Article 14 (2) Regulation (EU) 2019/943. If the ENTSO-E, a transmission operator or the regulatory authority notices a structural bottleneck, the member state must decide within six months together with the transmission system operator to "(a) establish national or multinational action plans in accordance with Article 15 Regulation (EU) 2019/943" or "(b) examine the bidding zone configuration or have it adjusted". If the decision is made to check the configuration of bidding areas, the relevant member states pass a unanimous resolution within six months. However, if the relevant member states cannot make a unanimous decision, the European Commission itself decides as an ultima ratio whether the configuration remains, or needs to be changed, Article 14 (8) Regulation (EU) 2019/943.[43]

This is the special feature of the new regulation. Until now, Regulation (EC) No. 714/2009, which has been recast by the above-mentioned Regulation, merely laid down general principles for bottleneck management in Article 16. Network bottlenecks should, therefore, be solved with non-discriminatory, market-oriented measures that send "effective economic signals to market participants and participating transmission system operators". The competence of the EU Commission to make an ultima ratio decision was not previously regulated there. Although Article 32 of the Regulation (EU) 2015/1222 (CACM Regulation) also contains a standard on the bidding zone configuration procedure, Article 32 (4) (b) and (c) only required a consultation, a workshop and a final decision by the transmission system operators and regulatory authorities to be submitted following the procedure. The only recourse that remained was to the general rule of Article 9 (11) of the CACM Regulation, which states that if the six-month period is not observed, ACER may adopt a decision on the proposals and methods submitted. It was, therefore, for the Member States to adopt or implement the joint proposal to amend or maintain the bidding zones.

---

43 Regulation 2019/943, Recital no. 31: "A Commission decision on the configuration of a bidding zone should be possible as a measure of last resort and should only amend the configuration of a bidding zone in those Member States which have opted to split the bidding zone or which have not reached the minimum level of the capacity."

5 A conflict with § 3a StromNZV?

With the Ordinance on Changes to the Electricity Network Access Ordinance,[44] (StromNZV), the German legislator created a legal basis for maintaining the historically uniform bidding zone in Germany.[45] It obliges transmission system operators not to endanger the uniform German bidding zone when carrying out trading transactions, cf. § 3a sentence 2 StromNZV. The reason behind the standardization was, on the one hand, to clarify who could decide on a division of the German electricity price zone. The legislator stated that it was unacceptable that such a division could be carried out without the involvement of public authorities.[46] In particular, transmission system operators were not to be left as the sole decision makers. On the other hand, network access should also be guaranteed uniformly throughout Germany, as the size and layout of the bidding zone would influence trade flows with neighbouring zones, including the liquidity of wholesale markets. Another intention was to avoid the bidding zone configuration procedure under Article 32 of the Regulation (EU) 2015/1222.[47]

The introduction of §3a StromNZV was strongly criticized in the literature: The distribution of bidding zones was already comprehensively regulated at a European level, due to the shared competence of the Member States and the EU under Article 4 (2) (i) TFEU, Germany had no legislative power.[48] This would be a violation of Article 194 TFEU in conjunction with Article 2 (2) TFEU. Even under substantive law, § 3a StromNZV was not compatible with the objectives of a harmonized EU internal electricity market. There is also a conflict with the free movement of goods under Article 34 TFEU. The legal

---

44 "Verordnung zur Änderung der Stromnetzzugangsverordnung" of 19 December 2017, BGBl. I 2017, 3988.
45 Wiegand Laubenstein, in: *Berliner Kommentar zum Energierecht*, ed. Franz Jürgen Säcker, Vol. 3, para. 3a StromNZV rec. 2.
46 Bundesministerium für Wirtschaft und Energie, "Referentenentwurf zur Änderung der Verordnung über den Zugang zu Elektrizitätsversorgungsnetzen", last modified 26 October 2017, https://www.bmwi.de/Redaktion/DE/Downloads/Gesetz/referentenentwurf-stromnzv.pdf?__blob=publicationFile&v=6 (accessed 4 June 2020).
47 Regulation (EU) 2015/1222 of establishing a guideline on capacity allocation and congestion management of 24 July 2015, OJ L 197, 25 July 2017, 24.
48 Carsten König and Max Baumgart, "Der EU-Binnenmarkt und die einheitliche Stromgebotszone in Deutschland", *Europäische Zeitschrift für Wirtschaftsrecht* (2018), 491, 492 et seq.

consequence would be an inapplicability of § 3a StromNZV.[49] With the, now, even more detailed regulations of the new Article 14 Regulation (EU) 2019/943, the arguments of the literature in view, gain strength. Notably, the declaration of competence that the European Commission can now decide as a last resort on the division of bidding zones is decisive regarding the discussion on legislative power. The Commission should only change the bidding zone configuration if the member states declared to divide up the zones, or if the minimum capacity[50] has not been reached by means of an action plan.[51] It is thus possible for the member states to independently prevent a final decision by the European Commission. This could be seen as a diplomatic success.[52] The regulation contains the principle that the member states should decide, while the European Commission can only do it exceptionally. For the concerned member state, the procedure offers the opportunity to vote in favour of an action plan. There seems to be some room for manoeuvre in this respect: It is not regulated what period should be covered, only that an action plan must be drawn up within six months.[53] It is becoming increasingly clear that the EU has made use of its legislative competence with the Regulation (EU) 2019/943 and has now, at least substantively, determined which parties will decide when, on the allocation of electricity price zones. At this point, at the latest, the EU had exercised

---

49 Prevailing opinion in the literature, see Walter Obwexer, in: *Europäisches Unionsrecht*, eds. Hans von der Groeben, Jürgen Schwarze and Armin Hatje (Baden-Baden: Nomos Verlag, 2015), Article 2 AEUV rec. 29; also Martin Nettesheim, in: *Das Recht der Europäischen Union*, eds. Eberhard Grabitz, Meinhard Hilf and Martin Nettesheim (München: C.H. Beck Verlag, 2019), Article 2 AEUV rec. 27.
50 For limits where the net coordinated transmission capacity approach is applied: 70 % of the transmission capacity in compliance with the operational safety limits and deduction of failure variants. For limits at which a load-flow-based approach is applied: 70 % of the internal and cross-zone critical network elements in compliance with the operational safety values and deduction of failure variants, see Article 16 (8) of Regulation 2019/943.
51 Regulation 2019/943, Recital no. 31.
52 Rainer Lütkehus, "Aufspaltung von Strompreiszonen nur als ultima ratio", *energate messenger*, 15 December 2017, https://www.energate-messenger.de/news/179547/aufspaltung-von-strompreiszonen-nur-als-ultima-ratio (accessed 14 November 2019).
53 Markus Kahles and Fabian Pause, "Überprüfung der einheitlichen deutschen Stromgebotszone nach der Elektrizitätsbinnenmarkt-Verordnung", *Würzburger Berichte zum Umweltenergierecht* (2019), 7.

its competence under Article 2 (2) TFEU, with the result that a blocking effect[54] has occurred which leads to the inapplicability of § 3a StromNZV. From a practical point of view, it is presumed that the minimum transmission capacity required by the new Internal Electricity Market Regulation can often not be met.[55] A permanent or temporary exemption is therefore not to be hoped for. Market participants should expect that one of the above-mentioned bodies will soon identify a structural bottleneck and seek a bidding zone configuration procedure.

Conclusions

The principal of electricity price zones may be helpful to simplify trade. Yet, the best solution for abolishing bottlenecks and the successful integration in the European market,[56] is the network extension. Market-splitting is not necessarily leading to more security of supply. Regarding the decision competence of the European Commission, applying the ultima-ratio-principle seems expedient. The member states, regulatory authorities, and transmission system operators, as well as the energy-intensive companies, are well-advised to put effort into solving the bottlenecks; otherwise, the Commission will take over. The new regulation with its specific effects on bidding zones is a sign of a constant change in an interdisciplinary field of law. But it also offers the opportunity to play a legal and economic role in shaping the European energy law landscape.

---

54 Martin Nettesheim, in: *Das Recht der Europäischen Union*, eds. Eberhard Grabitz, Meinhard Hilf and Martin Nettesheim, Article 2 AEUV rec. 27.
55 See the expert report Frontier Economics/Consentec, "Bedeutung von etablierten nationalen Gebotszonen für die Integration des europäischen Strommarkts", pp. 50 et seq.
56 Ibid.

Alexia Trokoudi & Emmanouil Chourmousis[*]

11 The General Court's Decision of 15 November 2018 (case T-793/14) for the annulment of the EC decision of 23 July 2014 (case S. 35980) on the Capacity Mechanism in the UK

Abstract: The chapter outlines the key points of the General Court's Decision of November 15, 2018 (case T-793/14), concerning the annulment of the EC Decision on the Capacity Mechanism which was to be implemented in the UK the following ten (10) years, to address market failures, i.e. reliability and the 'missing money' problem. The abovementioned Court's Decision was adopted in response to the appeal of two companies active in the Demand Side Response (DSR) market, and subsequently, the Court annulled the European Commission's decision approving such capacity mechanism; providing, henceforth, a 'compass' for the Commission's future assessments of the Member States capacity schemes and paving the way for more formal investigations, under Article 108 par. 2 of the TFEU and Article 6 of the Procedural Regulation. The Court's examination on the substance of the matter, focused on the aspects raised by the applicants pursuant to Article 4 of the Procedural Regulation, and more specifically that the proposed mechanism should have raised doubts as to its compatibility with the internal market. The Court found, inter alia, that the Commission lacked the required information to assess the DSR role in the UK market, potential infringements of the 'equal treatment' principle as regards the different treatment between the DSR operators and the electricity generators and the conditions of participation in the capacity market of the DSR. Therefore, the Commission should have had doubts about the mechanism's compatibility and conduct its own investigation. Following the annulment of the EC decision, the Commission launched an in-depth investigation into the 'UK capacity market'. Following the investigation, the Commission concluded that the UK Capacity

---

[*] Alexia Trokoudi is the Group Director of the Legal Division on Energy Regulation at Hellenic Petroleum S.A., and Emmanouil Chourmousis, LL.M., is Lawyer at Hellenic Petroleum S.A.

Market Scheme, for the period 2014–2024, is compliant with EU State Aid Rules for environmental protection and energy and does not disrupt competition.

Keywords: Capacity Mechanism, UK state aid rules, Compliance with EU State Aid Rules, annulment by General Court, doubts pursuant to the Procedural Directive, EU internal market, Demand Side Response

Introductory note

This chapter substantially depicts the points presented during the 4th Athens Conference on European Energy Law "Energy Arbitration and Judicial Dispute Settlement", which took place on September 27, 2019, with regards to the General Court's decision on the application for the annulment of the Commission's decision to approve the UK capacity. In our opinion, one can draw very useful conclusions by the said decision and this decision will be undoubtedly the compass for the Commission's future assessments of the Member States capacity schemes.

1 Commission's Decision C (2014) 5083[1] approving the UK Capacity Mechanism

1.1 The mechanism

Before proceeding to the legal considerations and the reasoning of the General Court's decision, it is of the utmost importance to make a few references to the annulled decision of the Commission. In accordance with the procedure stipulated in Article 108 par. 3 of the TFEU, the UK had notified as early as June 23, 2014 the capacity measures which it intended to implement for a period of 10 years. According to the UK, the mechanism is necessary due to market failures, i.e. reliability and the missing money problem. Those two market failures may be aggravated in the short and medium-term due to the expected closure of a fifth of the existing capacity in the next decade, and the reliability adequacy standard may reach critical levels as 2018/2019. The annual budget of the mechanism was estimated to rise to an amount between 0.9 billion GBP and 2.6 billion GBP. In essence, pursuant to the proposed mechanism, the UK

---

1 European Commission, State aid – United Kingdom Electricity market reform – Capacity market, C (2014) 5083 final of 23 July 2014, https://ec.europa.eu/competition/state_aid/cases/253240/253240_1579271_165_2.pdf (accessed 4 June 2020).

would organize centrally-managed auctions to procure the level of capacity required to ensure generation adequacy. The auctions would be open to electricity generators (existing and new), demand-side response operators and storage operators. The capacity providers would be entitled to a steady payment during the duration of the capacity agreement. The compensation is calculated for all successful participants as the last-accepted bid in the auction. Depending on the type of the generator, whether it is a new, refurbished or old plant, the duration of the capacity agreement issued by the UK system operator, under the name National Grid, may vary from 1 up to 15 years in order to ensure a level playing field. Most capacity providers would be granted 1-year-agreements, whereas refurbished plants would be eligible for up to 3 years and new plants for up to 15 years. The capacity providers are also subject to penalties in case they fail to deliver their undertaken obligations.

With regards to the beneficiaries of the mechanism, the prospective capacity providers may participate individually or in aggregate with other generating units in the auctions provided that their capacity is between 2 and 50 MW. The mechanism provides for exceptions to the participation in the auctions for those units, which already receive support or funding through other mechanisms (e.g. support for RES, small-scale feed-in Tariff, contracts for differences, etc.). The capacity, which could be provided from the interconnections by plants outside GB, was also exempted until 2015. The auctioned capacity would be determined annually on the basis of a reliability standard to be carried out by the system operator and the final decision on the amount of the capacity would be taken by the UK government. The main auctions are planned annually for delivery after four years, and in addition, year-ahead auctions would take place the year before the delivery of the main auctions for a capacity reserved from the four year-ahead capacity. Each auction would be a descending-clock, pay-as-clear auction. It should be noted that the capacity mechanism envisages the year-ahead auction as a better route for demand-side response (DSR) participants. Additionally, two transitional auctions limited to DSR were scheduled for the years 2015 and 2016. Finally, the UK licensed suppliers would provide for the financing of the capacity mechanism. In view of the above, the UK stated that the proposed mechanism is compatible both with the internal market pursuant to Article 107 par. 3 c TFEU and the Guidelines on State Aid for environmental protection and energy 2014–2020 (EEAG). In this respect, the UK argued that the concerns raised by a balancing services operator, an operator owning existing plants and by operators in the Demand Response market regarding the compatibility of the proposed mechanism with the internal market are unfounded.

1.2 Commission's assessment

Following the above, the Commission assessed the proposed mechanism. More specifically based on Article 107 par. 1 TFEU the Commission concluded that the proposed mechanism is imputable to the UK, confers an advantage on the undertakings of the electricity, which would receive compensation and as a result, it constitutes a State Aid. Furthermore, on examining its compatibility with the internal market based on the EEAG, the Commission found that the proposed measures contribute to an objective of common interest, are necessary, appropriate and proportional. Consequently, the Commission found the proposed mechanism compatible with EEAG and the internal market and decided not to raise objections.

1.3 The judgment of the EU General Court

On November 15, 2018, the EU General Court reached its verdict on the application of Tempus Energy Ltd and Tempus Energy Technology, both UK based companies active on the DSR market (hereinafter "Tempus" or "applicants")[2],[3]. The action (application) was filed on December 4, 2014 pursuant to Article 263 TFEU and sought the annulment of the abovementioned Commission decision not to raise objections to the UK capacity mechanism on the grounds that it is compatible with the internal market pursuant to Article 107 par. 3 c TFEU. In its pleadings to the Commission the applicants did not contest the adoption of the proposed mechanism as such but the Commission's assessments regarding DSR and the procedures for the participation of the DSR operators in the capacity market.

---

2 General Court, judgment of 15 November 2018, case T-793/14, ECLI:EU:T:2018:790, *Tempus Energy Ltd and Tempus Energy Technology Ltd v. European Commission*; "Action and main arguments brought on 4 December 2014", *InfoCuria Case-law*, http://curia.europa.eu/juris/document/document.jsf?text=&docid=162815&pageIndex=0&doclang=EN&mode=req&dir=&occ=first&part=1&cid=7203847 (accessed 4 June 2020).

3 General Court, judgment of 15 November 2018, case T-793/14, ECLI:EU:T:2018:790, *Tempus Energy Ltd and Tempus Energy Technology Ltd v. European Commission*.

2 Procedural steps before the Court – Admissibility of the action

After the finalization of its composition, the Court began its actions by inviting the Commission to submit the UK's notification of the proposed mechanism. Following the receipt of the notification of the measure, the Court informed the UK and the Commission of its intention to include a number of extracts from the notification in the case-file. At the same time, the Court invited the UK and the Commission to submit their observations regarding the disclosure of the said extracts. Further, the Court invited all parties to the procedure to answer a series of questions. Those questions were related to the Guidelines (EEAG) and the information available when the contested decision was issued regarding the DSR and its possible technology developments and their impact on UK security of supply. In addition, Tempus was invited to comment on Commission's statements regarding the cost recovery method and on the other hand the Commission was asked to indicate the criteria which formed the basis of its assessment regarding the financing needs of the DSR operators. As regards to the admissibility of the action, the Commission argued that the action is partly inadmissible and more specifically as regards to the part of the application that does not relate to the defence of Tempus' procedural rights. After the hearing, the Court rejected the Commission's argument regarding the admissibility of the action.

3 Court's legal considerations on the substance of the case

3.1 Concept of doubts

Tempus' first and foremost allegation was that the Commission following its preliminary review of the proposed mechanism could not draw the conclusion, taking into account the information available at the time, that the proposed mechanism did not raise doubts as to its compatibility with the internal market. In order to make its point clear, Tempus presented seven (7) aspects that give rise to such doubts pursuant to Article 4 of Regulation 659/1999. As a result, Tempus alleged that the contested decision privileges generation over DSR in a discriminatory and disproportionate manner. In Tempus' view the Commission should have initiated a formal investigation procedure pursuant to Article 108 par. 2 TFEU, which, by the issuance of the contested decision, is infringed. The Court proceeded to the analysis of the relevant provisions and case law regarding the concept of "doubts" pursuant to Articles 4 par. 3

and 4 of Regulation 659/1999, pursuant to which Tempus argument, that a formal investigation procedure should have been conducted, was founded. In this respect, the Court concluded that for the establishment of the existence of doubts (Article 4 par. 4 of Regulation) it is sufficient that Tempus show that the Commission has not researched and examined, thoroughly and impartially, all of the relevant information or it has failed to take them into consideration so as to eliminate all doubt about the compatibility of the mechanism. It should be noted that the Commission should be judged by the information available on the date that the decision was adopted.

3.2 Length of discussions between the Commission and the UK

Tempus alleged that the significant length of the negotiations between the Commission and the UK prior to the notification of the mechanism signals an indication that may establish that there were doubts. The Commission dismissed the said allegation and argued that only the date of the notification matters. Furthermore, given that the procedure of the preliminary assessment lasted less than two months, it means that no doubts have arisen. The Court itself considered on the basis of the facts of the case that the notified mechanism is a significant, complex and novel measure in view of the very high amounts of aid involved, the complex implementation procedures and most importantly the fact that it is the first mechanism assessed in the light of the Guidelines on State Aid (EEAG). In view of the above, the Court considered important that the length and the content of the contacts between the UK and the Commission prior to the notification shall be taken into consideration. In fact, the Court makes reference to the Best Practices Code in the conduct of State Aid proceedings and maintains that the pre-notification period, which includes the contact between the Commission and UK, exceeded significantly the general rule of the two-month period provided in such code and lasted more than a year. Furthermore, despite the novelty and complexity of the measure as evidenced by the 3 rounds of questions by the Commission during the pre-notification period and Commission's pending questions specifically on the incentive effect, proportionality and the existence of potential discrimination did not stop the Commission to declare in a letter to the UK the proposed mechanism as *prima facie* compatible with the TFEU and the Guidelines. Such a conclusion was drawn before the preliminary examination of the mechanism pursuant to Regulation 659/1999. In addition, three types of operators challenging the proposed mechanism had already intervened before the Commission. Finally, the Court points out that the DSR role in the mechanism does not seem to have been independently analysed by the Commission. In view of the above, the Court is of the view that the contested decision should be examined so as to assess whether there are indications that the Commission should have had doubts.

3.3 Assessment of the DSR role in the UK capacity market

Moving to the examination of the arguments on the substance of the case at issue, the Court notes, after evaluating the allegations of all parties, that the key question is whether the role which DSR is capable of playing within the UK capacity was properly evaluated by the Commission. The Court, in order to answer this question, proceeded to the analysis of the role of DSR and its potential and concluded that on the basis of the information available to the Commission at the time of the preliminary examination and especially the Panel of Technical Experts (PTE) report should have led the Commission to initiate its own analysis of the role and potential of the DSR. In fact, the PTE report expressed its concerns that the demand side was not yet understood as well as conventional generation, and as a result, the role and the potential of the DSR in the capacity market were not obvious and clear. Furthermore, the PTE recommended that the initiation of a program to fully research the potential of the DSR in view of the upcoming auctions. In this respect, the Court also referred to the need to assess the role of the DSR on the basis of the Guidelines (EEAG). In view of the above, the Commission should not have been satisfied by the information available on DSR and as a result, there were doubts as to the compatibility of the scheme with the internal market.

3.4 Discriminatory treatment of DSR in the UK capacity market

The final line of Tempus' argumentation concerned an alleged discriminatory treatment of DSR within the capacity market in contrast with the more favourable treatment to generators. Tempus particularized such a claim with reference to four points proving such argument, namely the length of capacity contracts, the capacity market's cost recovery method, the conditions for the participation in the auctions and the lack of additional remuneration of the DSR operators. The Court evaluated each one of the above lines of argumentation. More specifically and on the subject of the length of capacity contracts, the Court considered whether the difference in treatment between the DSR operators (entitled to 1-year contracts) and the generators (some categories were entitled to multiyear contracts – up to 3-year and up to 15-year contracts) was justified. Again, the issue at hand was whether the Commission should have initiated its own formal investigation or not. The Court concluded that once again the Commission did not have the required information to assess in detail the capital expenditure and financial needs of the DSR operators, although these financial criteria were the basis for the allowance of multiyear contracts to generation. As a result, without such information, the Commission could not check the potential infringement of the equal treatment principle, and it should have doubts on the compatibility of the mechanism with the internal market. With regards to the second argument on the cost recovery method of the mechanism, Tempus argued that the adopted method does not incentivize consumers to reduce their consumption and as a result, more capacity is needed, which results in the increased cost of the mechanism. On the other hand, the Commission dismisses such an argument. The Court noted that the Guidelines describe an aid measure as proportionate, only if its amount is limited to the minimum needed to reach its objective. In essence the UK, only after the public consultation, changed its initial decision and adopted a cost recovery method based on electricity consumption between 16:00 and 19:00 during winter weekdays (from November until February). Given the fact that the cost recovery method influences the volume of capacity, the Commission should have investigated whether the new method had the same incentive to the consumers' consumption and the development of DSR. Without all the information, such examination was not possible, which pointed to an indication that there were doubts as to the compatibility of the scheme. The next argument of Tempus concerns the conditions of participation in the capacity market of the DSR and especially whether it is adequately incentivized pursuant to the Guidelines without having to face economic or other barriers. The Court found that, although DSR can participate both in the transitional and the enduring auctions contrary to Tempus claims, the Commission should have had doubts as to the size of the incentive for the participation of the DSR in the year ahead (T-1) auctions, which represent the better route to the auctions of the DSR, in view of the limited capacity reserved for such auctions. Finally, the Court concluded that the participation threshold of 2 MWs de-minimis should have been considered by the Commission as raising doubts. In view of the above, and following the examination of Tempus argumentation on the matter, the Court viewed that the Commission should have had doubts as to the compatibility of the UK mechanism with the internal market. The final argument assessed by the Court was the lack of additional remuneration to the DSR operators for the alleged savings in the amount of electricity lost during transmission and distribution. The Court evaluated the said argument under the Guidelines and ended up rejecting it, due to the fact that pursuant to the Guidelines the aid measures (such as the UK capacity mechanism) remunerate only the service of pure availability, and such remuneration does not include any remuneration for the sale of electricity (remuneration per MWh sold).

4 The procedural events following the annulment of the UK capacity mechanism

The annulment of the abovementioned decision resulted in the opening of an in-depth investigation into the UK capacity market by the Commission to determine whether such mechanism is in line with EU State aid rules and to comply with the court's decision.[4] Simultaneously the Commission has appealed the General Court's judgment to the European Court of Justice.[5] Following the completion of the in-depth investigation, the Commission has confirmed in its recent decision,[6] that the UK Capacity Market scheme covering the period 2014–2024 complies with EU State aid rules. Moreover, the Commission announced that it did not find any evidence that the scheme would put demand response operators or any other capacity providers at a disadvantage with respect to their participation in the scheme and such mechanism is necessary for the security of supply in the UK and does not distort competition in the energy market.[7]

---

4 European Commission, "State aid: Commission opens in-depth investigation into British Capacity Market scheme", Press release IP/19/1348 of 21 February 20 19, https://ec.europa.eu/commission/presscorner/detail/en/IP_19_1348 (accessed 4 June 2020).
5 Appeal brought on January 25, 2019 by the European Commission against the judgment of the General Court (Third Chamber, Extended Composition) delivered on 15 November 2018 in case T-793/14, *Tempus Energy and Tempus Energy Technology v. Commission*, case C-57/19 P, http://curia.europa.eu/juris/document/document.jsf?text=&docid=213445&pageIndex=0&doclang=en&mode=lst&dir=&occ=first&part=1&cid=7205105 (accessed 4 June 2020).
6 European Commission, Decision on The Aid Scheme SA.35980 – 2019/C United Kingdom – Electricity Market Reform: Capacity Mechanism, C(2019) 7610 final of 24 October 2019, https://ec.europa.eu/competition/state_aid/cases1/201945/278880_2105752_352_2.pdf (accessed 4 June 2020).
7 European Commission, "State aid: Commission approves the British Capacity Market scheme", Press release IP/19/6152 of 24 October 2019, https://ec.europa.eu/commission/presscorner/detail/en/ip_19_6152 (accessed 4 June 2020).

Conclusions

The Court's decision on the annulment of the UK capacity mechanism enables us to draw useful conclusions but also raises questions with regards to the future approvals of capacity schemes. It should be underlined that the European Court did not decide whether the contested mechanism is compatible with the internal market but the annulment of the Commission's decision on grounds deriving solely from the assessment procedure of the UK's notification. In light of the foregoing, it is evident that from now on, the Member States wishing to initiate a capacity mechanism must accompany their formal notifications with technical evidence so as to satisfy the procedural regulation provisions regarding the elimination of doubts. However, it is not obvious what the required detail of an impartial and well-reasoned technical evidence is, in order for such evidence to be considered complete and satisfactory before the Court in a similar future case. On the other hand, the Commission will have to double-check the assumptions and information provided by the Member States in order not to initiate a formal investigation. This case paves the way for more formal investigations under 108 par. 2 of the TFEU and Article 6 of the Procedural Regulation. More specifically, where there is an absence of independent and impartial evidence or lack of information, as in the case at hand, the European Court expects that the Commission will conduct its own assessments and use its powers as provided in the formal investigation procedure. Finally, the decision has made it clear that a prolonged pre-notification period leads to the conclusion that there are doubts about the compatibility of the proposed mechanism. In view of such notion, it remains to be seen whether the Commission will entertain in the future lengthy pre-notification procedures for similar cases or future pre-notification periods will be limited both in time and in subject.

Dr. Maria Lykidi[*]

12 Arbitration in the construction of nuclear power plants

Abstract: Arbitration in the nuclear sector is related mostly to the construction of nuclear reactors, primarily due to the high complexity that characterizes the projects of nuclear power plants from a technical and economical perspective as well as the requirements of nuclear regulation and their impact on the project schedule. This chapter focuses on the case of EPR which is a new generation nuclear reactor that embodies important improvements with respect to the existing nuclear reactors. Specific cases that could constitute possible candidates for construction arbitration are presented. The chapter examines the argumentation in each case as well as the final outcome. Special attention is also given to projects of extension of the lifetime of nuclear power plants and the position of the European Court of Justice regarding the necessity of implementation of environmental impact assessment studies before their realisation.

Keywords: Nuclear sector, construction arbitration, nuclear regulation, EPR, start-up delays, investment costs, lifetime extension, environmental impact

1 Main issues resulting in arbitration in the nuclear sector

International arbitration in the nuclear sector is related mostly with nuclear reactors' construction arbitration to which the nuclear industry and environmental associations pay particular attention. It is due to the high complexity that characterizes the projects of nuclear power plants from a technical and economic point of view. For this reason, such arbitration procedures are often long-running and involve large teams of legal and technical nuclear experts. The key points that could be in issue in a construction dispute in the nuclear sector relate to the licensing risk, the risk of constructing in a nuclear environment, and nuclear liability. The licensing risk is mainly related to the possibility of obtaining the license in order to proceed with a nuclear power plant (NPP) project, the conditions indicated in the license and the time until the license

---

[*] Dr. Maria Lykidi is an Energy markets and Modelling Expert and a Research Associate at the Hellenic Energy Regulation Institute.

is obtained.[1] In general, licensing procedure contains several milestones such as early site permits, design certification, the construction license, operating license, etc. Each milestone is important for the timeline and the financial cost of the project. Any delayed achievement of these milestones can be the reason for starting a construction arbitration process where the supplier including e.g. key component vendors or subcontractors, are claiming additional time or money, or the owner is seeking to deduct liquidated damages. For this reason, the scientific opinion of nuclear experts plays a crucial role in the evolution and outcome of the arbitration since they have the expertise to analyse the licensing procedure and say whether the cause of delays can be justified or not, or/and provide their opinion as to the probability of new or differentiated preconditions imposed by the regulator. Another parameter that needs to be taken into account is the nuclear regulator and its role in the implementation of an NPP project, given that it has a high safety classification with respect to other construction projects. The nuclear regulator will be interested (and potentially intervene) in the entire supply chain, supervising each phase of the project, from basic through to detailed design, procurement, manufacturing, factory testing, construction and installation through to commissioning. In view of the high responsibility of the nuclear regulator for supervising and intervening in nuclear works to ensure nuclear safety and the fact that regulators do not bear the cost of their interventions, they can proceed with all the necessary changes, interventions and additional or different requirements. Their "behaviour" can have a considerable impact on the project schedule and cost, giving rise to claims in arbitration as to who must bear the financial consequences between the owner and supplier through the supply chain. We keep in mind that every NPP project has its peculiarities that differentiate it from other NPP projects and a different implementation timing. For example, even if a first-of-a-kind NPP is constructed (e.g. the case of EPR), additional NPPs of the same design still need to be further developed; the supplier, the owner and the working staff can be different; the country, the nuclear regulator as well as the regulation itself may also be different. Therefore, no NPP project can be used as a reference that other NPP projects may follow as it can create a considerable technical and legal argument during an arbitration. The risk of constructing in a nuclear environment consists of the risk entailed in the supplier's work during the

---

1 "Liability for Nuclear Damage", *World Nuclear Association*, August 2018, https://www.world-nuclear.org/information-library/safety-and-security/safety-of-plants/liability-for-nuclear-damage.aspx (accessed 4 June 2020).

different phases of implementation of a nuclear project. In the nuclear sector, an NPP construction project needs to be conformed to many strict regulations that limit the ability of the supplier to work "at risk" since it cannot take any action without prior formal approval.[2] Particularly, if we take into consideration the safety requirements imposed by the ASN regarding specific features of a NPP. It is very difficult to avoid delays leading to extra costs during the realization of the project in a "zero risk" context given also that the owner can require opening up an investigation of the works at the supplier's cost in order to check that the work done is compliant to the contractual specifications.

In the case that the investigations show that the work is non-compliant, the supplier may have to rework at the project's timetable and cost. Consequently, any delay in approval by the owner or the regulator can have a significant impact on the project schedule (and create substantial additional cost). The proper assessment of the full impact of all possible delays on an NPP project has to be taken into account by the experts that are invited to provide their opinion in an arbitration process. Similarly, another point that needs to be highlighted is the limitations in the management of the internal processes necessary for the proper implementation of the entire supply chain of the nuclear project because of the existence of tight nuclear regulations. For example, in the phase of construction of an NPP, where there are limitations regarding the work that is realized, the supplier may not be allowed to adopt specific approaches or amendments and developments (e.g. on the design of the construction site, on the project team work practices, etc.) or implement suitable solutions where and when an issue (e.g. welding concerns,[3] claims for loss of productivity of the work teams, etc.) arises without formal approved construction drawings. Thus, a small construction issue can cause important delays to the construction site works while the design is reworked, and approvals are processed. Therefore, in both cases, the experts and counsel must be able to fully comprehend these components of constructing in a nuclear environment and be able to explain to the tribunal how the regulatory environment impacts on the owner and supply chain, and who bears the risk. The nuclear liability in the case of nuclear damage is the

---

2   Stavros Brekoulakis and David Brynmor Thomas, "The Guide to Construction Arbitration – Second Edition" (London: Law Business Research Ltd, 2018).

3   Under many nuclear regimes, the approval of welders is highly regulated, with different welds requiring different certifications and approvals. As a result, the supplier is unable to reallocate welders freely, and must match specific welders to specific tasks.

third key point that deserves attention in the nuclear sector.[4] Liabilities for nuclear damage are under the jurisdiction of the national legislation as well as of international conventions,[5] given that the consequences of a nuclear incident can be spread in a large geographical area. According to these conventions, liability for nuclear damage results exclusively from the operator of the nuclear installation who has also financial liability for the incident subject to certain limits. Moreover, it is under the exclusive jurisdiction of the national court of the country where the nuclear incident took place (normally the location of the NPP) to decide on compensation claims or any other nuclear damage claims. Regarding the nuclear liability in the nuclear constructions, the supplier who is responsible for the supply chain of an NPP project should not face direct claims in relation to nuclear damage in view of the basic principles of the conventions mentioned above. However, if the contract of the construction of the NPP that exists between the operator and the supplier allows the operator to pass liability for nuclear damage down to the supplier, then nuclear damage claims can still appear in construction arbitrations.

2 Construction arbitration in the nuclear sector – The case of EPR

The European Pressurized Reactor (EPR) which is an evolution of the Pressurized Water Reactor (PWR) is an example of III+ generation advanced nuclear reactor,[6] which is designed to incorporate significant improvements.

---

4 European Commission, "State aid: Commission concludes modified UK measures for Hinkley Point nuclear power plant are compatible with EU rules", Press release IP/14/1093 of 8 October 2014, https://ec.europa.eu/commission/presscorner/detail/en/IP_14_1093 (accessed 4 June 2020).

5 The four most significant international conventions are the following: (i) Paris Convention on Third Party Liability in the Field of Nuclear Energy 1960, (ii) Brussels Supplementary Convention 1963 to the Paris Convention, (iii) Vienna Convention on Civil Liability for Nuclear Damage 1963, (iv) Joint Protocol Relating to the Application of the Vienna Convention and the Paris Convention 1988.

6 Nuclear reactors of modern design (third generation and its evolution III+) are capable of flexible operation which is very different from baseload operation; in flexible operation, the amount of electricity supplied by a generating system at any given time (load) follows the predicted evolution of energy demand (daily and seasonal). A nuclear unit can vary its capacity level between the nominal capacity and the technical minimum. The minimum requirements for the maneuverability of modern reactors are defined by the utilities requirements that are based on the requirements of the grid operators.

These are related to the reliability of safety systems and the technology instrumentation and control in order to eliminate or mitigate operational human mistakes which became crucial particularly after the Fukushima accident in 2011. Additionally, important updates have been realized in the operation which benefits from the accumulated experience of the previous nuclear reactor design, construction, commercialization and operation. The conception of its design resulted initially from the cooperation of French and German makers of European nuclear power plants (Framatome and Siemens).[7] At the end of 1991, Electricité de France (EDF) and the majority of German electrical operators joined together their own development programs in order to design a new PWR reactor being the EPR design. The goal was to achieve the optimal design through the merge of the characteristics of French and German designs. Therefore, a safer, more efficient and more competitive nuclear reactor was conceived based on: (i) the high level of operational experience coming from the construction of an important number of nuclear reactors from Framatome and Siemens contributed to the creation of an evolutionary design with special focus on safety systems features, (ii) another feature being the competitiveness of the new reactor in terms of costs with respect to alternative generation technologies. EPR ensures lower power generation costs with respect to other modern nuclear power reactors. The economic viability of the EPR is also provided through some key features such as a greater electrical power output (1600 MW), a more efficient use of the nuclear fuel and an increased availability preventing long outages. Specifically, the EPR design philosophy has the main objectives which are presented as follows:

- To reduce core damage frequency.[8]
- To decrease the level of radioactive waste per MWh produced.
- To mitigate or eliminate severe accidents.
- To protect critical systems from external events.

---

7   Jorge Morales Pedraza, "Small Modular Reactors for Electricity Generation" (Basel: Springer International Publishing, 2017).
8   A nuclear reactor core is a key component of a nuclear reactor which is the heart of a nuclear power plant. Nuclear reactors contain and control nuclear chain reactions that take place in the core and produce heat through a physical process called fission. That heat is used to make steam that spins a turbine to create electricity, "NUCLEAR 101: How Does a Nuclear Reactor Work?", energy.gov, 19 May 2020, https://www.energy.gov/ne/articles/nuclear-101-how-does-nuclear-reactor-work (accessed 4 June 2020).

- To improve the availability of the nuclear power plants and thus, their efficiency.
- To achieve a reduction of the consumption of nuclear fuel per MWh produced.
- To increase the operating lifetime of the nuclear reactor.
- To ameliorate the management of the nuclear fuel cycle.[9]
- To provide greater flexibility,[10] in order to accommodate load following operation if the grid system operator asks them to, in order to balance supply and demand and therefore ensure the stability of the electricity grid.[11]

---

9   The nuclear fuel cycle is the series of industrial processes which involve the production of electricity from uranium in nuclear power reactors, "The Nuclear Fuel Cycle", *World Nuclear Association,* March 2017, https://www.world-nuclear.org/information-library/nuclear-fuel-cycle/introduction/nuclear-fuel-cycle-overview.aspx (accessed 4 June 2020).
10 Economically, in view of the high fixed costs of nuclear, operators would want their nuclear plants to run constantly at full capacity in order to amortize their fixed costs. Furthermore, within the electricity market, the low variable costs, place nuclear at the bottom of the merit order behind the fossil fuel technologies (i.e. coal, gas, oil, etc.) making nuclear "economically suitable" to operate at full-load. Consequently, we deduce that nuclear operators would not prefer their units to take part in load-following operations but to be operated as baseload units. However, it will become necessary and inevitable for nuclear generating units to be able to operate flexibly under one or more of the conditions described below: (i) large percentage of nuclear generation capacity; (ii) growth in renewable or non-dispatchable production. If there is an important share of intermittent and nuclear power sources on the same electricity grid, the nuclear power plants must be capable of operating in load following mode to balance the fluctuations of total power generation; (iii) large nuclear generating units in a small electrical system; (iv) transmission system constraints since the capacity of the transmission network to which the NPP is connected may be limited because of e.g. a growth in generation, such as renewable generation, connected near the NPP; (v) constraints on non-nuclear (thermal) generating units because of increasingly strict environmental legislation in order to e.g. reduce CO_2 emissions and dependence from fossil fuels; (vi) changes in electricity market rules. A number of countries have deregulated their public electricity supply system and more countries are considering deregulation. The technical or commercial rules in the deregulated market may require all generating units to be treated similarly, and hence, require all generating units to have at least a defined minimum capability to operate flexibly.
11 Maria Lykidi and Pascal Gourdel, "The optimal short-term management of flexible nuclear plants in a competitive electricity system as a case of competition with reservoir" (Paris: Documents de travail du Centre d'Economie de la Sorbonne, 2014).

The nuclear projects of construction of an EPR that could be potential candidates for arbitration due to serious unforeseen delays in the construction of nuclear reactors and downward adjustments of the schedule for their commercial operation which led to an under-implementation (i.e. an over-budgeting) are hereby presented.[12]

2.1 Olkiluoto (OL3) – Finland

The construction of the EPR nuclear reactor started in 2005 with an initial scheduling to go online in 2009. It resulted from the cooperation between French AREVA and German Siemens via their common subsidiary AREVA NP for Finnish operator TVO (Teollisuuden Voima Oyj). The initial estimated cost was about 3.7 billion euros, but due to many delays the costs have been severely updated upwards several times. In 2015, the Finnish operator TVO stated in its annual report that the schedule had been updated by the supplier, and therefore regular electricity production would commence at the end of 2018. However, this did not happen, mainly because of difficulties on management and technical specifications, quality-control issues as well as communication and control complications due to the different nationalities of workers and sub-contractors. Additional delays have been noticed since then and a "regular electricity generation" was expected in September 2019. In view of the initial prediction to completion (56 months), its total construction time to operation on the current schedule has been about 3 times up, and operation ten years behind schedule. Due to budget increase and delays in the construction the Finland operator TVO and AREVA openly fought over financial liability leading to an arbitration process. In particular, they have sued each other, and the claims included delayed payments by TVO, penalty interest, and alleged loss of profit.

However, in 2016, AREVA NP called-off arbitration negotiations for a settlement with TVO. Finally, in March 2018, the parties announced that they had reached an agreement on the completion of OL3, including all related disputes. The agreement ensured financial compensation to TVO by AREVA in relation to costs and losses caused by the delays, commitment by AREVA that there were sufficient funds for completion of OL3 including the creation of

---

12 Jorge Morales Pedraza, "Small Modular Reactors for Electricity Generation" (Basel: Springer International Publishing, 2017); Mycle Schneider and Antony Froggatt, "The World Nuclear Industry Status Report 2018", *World Nuclear Industry Status Report*, 6 September 2018, https://www.worldnuclearreport.org/The-World-Nuclear-Industry-Status-Report-2018-HTML.html (accessed 4 June 2020).

a "trust mechanism" funded by AREVA to secure the financing of the costs of completion of the project as well as payment of a penalty to TVO in the event that AREVA fails to complete the project by the end of 2019. It should also be noted that, the International Chamber of Commerce (ICC) arbitration process, which had been considering liability over OL3, issued a partial ruling in TVO's favour in July 2017. Nevertheless, all parties announced their withdrawal from ICC proceedings with the settlement agreement in March 2018. Currently, the AREVA-Siemens consortium has informed the Finnish utility TVO that commissioning of the OL3 EPR is running six weeks behind schedule. In an updated schedule provided to TVO in July of the same year, the supplier said nuclear fuel was to be loaded into the reactor in January 2020, the first connection to the grid to take place in April 2020, and the start of regular electricity production was to take place in July 2020.[13]

2.2 Flamanville – France

The project of construction of an EPR nuclear reactor in Flamanville aims to maintain nuclear competence given that the French nuclear set is ageing (its average age was 33.5 years by mid-2018 in a lifetime of 40 years for some reactors) and no new reactor commissioning is realized together with the increasing development of RES. Initially, EDF said France's first EPR would cost 3.3 billion euros and start commercial operations in 2012, after construction lasting 54 months. However, both the duration of the construction and the construction costs have raised significantly due to a number of delays of several months that occurred for technical and safety reasons (e.g. detailed-design issues and quality-control problems) indicated by ASN (Autorité de Sûreté Nucléaire). In July 2018, EDF revised the schedule of construction and the loading of nuclear fuel was scheduled for the 4th quarter in 2019 being at least 7.5 years behind schedule and the targeted construction costs have reached the 10.9 billion i.e. three times its original estimate. Clearly, Flamanville's project is one of the projects that are many years behind schedule and billions over budget. Nevertheless, we do not find any cases of arbitration as we do in the project of Olkiluoto since EDF is a state-owned company that operates all 58 nuclear reactors being the main producer of electricity in France, with a total capacity of 63 GW.

---

13 "Olkiluoto EPR start-up delayed by final checks", *World Nuclear News*, 8 November 2019, https://www.world-nuclear-news.org/Articles/Delay-in-final-checks-puts-back-Olkiluoto-EPR-star (accessed 4 June 2020).

2.3 Hinkley Point – United Kingdom

Hinkley Point C, in the UK, is a project planned in the general context of plans for plant life-extension and nuclear new-build which remain major areas of investment in a few specific countries. Persistence to projects like this is observed despite increasing costs over original cost estimates, a number of technical difficulties and demands for government financial concessions and guarantees. In 2013, planning consent for Hinkley Point C nuclear power plant was given and at the same year EDF agreed with the UK government after a long period of negotiations. However, the European Commission has opened an in-depth investigation to examine whether UK plans to subsidize the construction and operation of a new nuclear power plant (EPR) at Hinkley Point involved state aid in the meaning of the EU rules,[14] and if yes, whether the plans were compatible with EU state aid rules.[15] The Commission had doubts that the project would need supportive market mechanisms in order to be viable.

According to the Commission Vice-President *Joaquín Almunia* in charge of competition policy during that period, the mechanism proposed by UK whose purpose was to attract investment in nuclear energy was complex and of an unprecedented nature and scale. The measure notified by UK was nothing else than the establishment of a feed-in-tariff mechanism ensuring that the nuclear operator would receive a stable revenue for a period of 35 years despite the volatility of the wholesale electricity price. In this way, the nuclear operator could avoid its exposure to market risk as it is the case for RES (Renewable Energy Sources). Moreover, the operator would benefit from a state guarantee covering any debt which the operator will seek to obtain on financial markets to fund the construction of the plant. So, the main question for Commission was if the

---

14 Public interventions in favour of companies can be considered free of state aid within the meaning of EU rules when they are made on terms that a private operator would have accepted under market conditions (the Market Economy Investor Principle – MEIP). If the MEIP is not respected, the public interventions involve state aid within the meaning of EU rules (Article 107 of the Treaty on the Functioning of the European Union – TFEU), because they confer an economic advantage on the beneficiary that its competitors do not have. The Commission then proceeds to assess whether such aid can be found compatible with the common EU rules that allow certain categories of aid.

15 European Commission, "State aid: Commission opens in-depth investigation into UK measures supporting nuclear energy", Press release IP/13/1277 of 18 December 2013, https://ec.europa.eu/commission/presscorner/detail/en/IP_13_1277 (accessed 4 June 2020).

project of Hinkley Point C could stand alone in the market without any state interventions or not.

At this point, some derogations need to be highlighted, which had been raised during that period.[16] In nuclear energy, there are no specific instruments that allow for support to get clearance by the European Commission (EC) as we have for RES (e.g. the General Block Exemption Regulation and the Guidelines for Environmental Aid). Consequently, this kind of state aid would have to be notified to the EC and the EC would have to assess whether it could be "cleared" as being compatible with the internal market. Generally, in order for state aid to be justifiable it would need to be aimed at a common interest and the appropriate instrument, have an incentive effect and be proportionate to the objective. Several derogations had been taken into account in the case of a complex generation technology such as the nuclear technology. Firstly, the transparency of the supportive mechanism that requires a clear and accurate evaluation and explanation of the total costs incurred by the state to avoid overcompensation. Nuclear is a special case given the magnitude and the incertitude that characterizes the investment costs, i.e. the costs involved prior to the electricity generation.

Starting from the costs for the import of nuclear material and the construction of plants to the costs for the disposal of nuclear waste, the necessary insurance amounts to cover the costs in case of a nuclear disaster as well as the costs due to tighter regulations particularly after the Fukushima accident, nuclear energy can be seen as an expensive technology compared to other technologies. Additionally, the construction of a "first-of-a-kind" nuclear reactor such as the EPR may hide significant financial risks as the examples of nuclear projects in Flamanville and Olkiluoto have shown. Nuclear costs are likely to be characterized as unclear and uncertain and thus, no cost calculation could occur for the grant of state aid. Secondly, the impact of the above-mentioned costs on the price per KWh of nuclear production could lead to an increase which makes nuclear less attractive and competitive versus gas power plants and RES. A third point is the characterization or not of nuclear technology as an infant technology, i.e. a new technology which needs support to enter the market

---

16 European Commission, "State aid: Commission concludes modified UK measures for Hinkley Point nuclear power plant are compatible with EU rules", Press release IP/14/1093 of 8 October 2014, https://ec.europa.eu/commission/presscorner/detail/en/IP_14_1093; Steve Thomas and Dörte Fouquet, "The new UK nuclear power programme – a FIT for nuclear and a blueprint for illegal state aid?", 2013, https://www.nuclearconsult.com/docs/Turmes_report.pdf (accessed 4 June 2020).

successfully. Some argued that EPR's technology is not a new or infant technology similar to "green technologies" including e.g. new electricity storage systems because even if it is a new generation-type reactor used for energy production, the process of generating energy would essentially remain the same.

Other derogations were related to: (i) the necessity of the aid given that the uncertainty in nuclear costs evaluation could lead to a feed-in-tariff supporting scheme that ensures higher profits for EDF than these obtained through the market, (ii) proportionality of the aid and non-discrimination. The EPR is designed to supply important amounts of generation capacity. Supporting such a technology and ensuring that aid is indeed proportionate to the objective pursued may have negative effects regarding the development and participation of other technologies in the national energy mix which may also lead to a discriminatory behaviour towards nuclear energy, (iii) the effect on competition and trade contrary to the common interest. Granting a state aid in favour of Hinkley Point C nuclear project through the feed-in-tariff supporting mechanism may cause distortions on competition opposing to the common interest since EDF's market position could be significantly reinforced in UK's energy market. Eventually, after 11 months of investigation, the EC concluded that the revised UK measures are compatible with EU rules and approved a UK support package in 2014.[17] UK's authorities proved that Hinkley Point C nuclear project would not be able to obtain the necessary financing due to its unprecedented nature and scale. Hence, the main revisions that UK undertook resulted in a state aid that remained proportionate to the objective pursued, avoiding however any undue distortions of competition in the European single market and reducing UK citizens' financial contribution to the project through: (i) the increase of the state guarantee fee that the operator would have paid to the UK Treasury which decreases the subsidy granted by the state, (ii) a gain-share mechanism that will be in place for the entire lifetime of the project, namely 60 years ensuring benefits to UK consumers. Nevertheless, Austria and Luxembourg, on June 2015, launched their appeal in the General Court of the European Union, challenging the EC's decision.[18] The main arguments

---

17 European Commission, "State aid: Commission concludes modified UK measures for Hinkley Point nuclear power plant are compatible with EU rules", Press release IP/14/1093 of 8 October 2014, op. cit.
18 "The General Court confirms the decision by which the Commission approved the aid provided by the UK in favour of the Hinkley Point C nuclear power station", General Court, judgment of 12 July 2018, case T-356/15, ECLI:EU:T:2018:439, *Austria v. Commission, Luxembourg*.

concerned the following points: (i) promotion of nuclear electricity does not constitute an objective of "common" interest capable of justifying aid, (ii) the technology used at Hinkley Point C is not new, (iii) absence of an incentive effect for the intervention of the UK, (iv) the proportionality of the aid in question, (v) the evaluation, by the Commission, of the positive and negative effects of the measures at issue is erroneous, (vi) the characterization of the measures as investment aid or as operating aid and (vii) UK ought to have launched a tendering procedure for this project.

Eventually, the General Court turned down the action brought by Austria, supported by Luxembourg and confirmed the decision by which the EC had approved the state aid in favour of the Hinkley Point C nuclear power station. The main points that responded to the above arguments by Austria were presented accordingly by the General Court as follows: (i) the objective pursued by a Member State must be one of public interest without meaning that it necessitates to be an interest of all the Member States or of a majority of them, (ii) the rules on state aid and the Euratom Treaty do not require the existence of technological innovation to be established within the nuclear project, (iii) state intervention is necessary in order to protect investments in new nuclear generating capacities from financial risks to which nuclear technology is subject given the absence of appropriate market-based financial instruments that could be used for hedging, (iv) impossibility to prove equivalence between building the nuclear generation capacity of Hinkley Point C and a comparable amount of wind generation capacity within the same time period because of the intermittent nature of RES, (v) UK has the right to determine its national energy mix and decide whether to maintain and increase or not nuclear capacity, (vi) irrespectively of the characterization of the aid measures at issue, they are appropriate and necessary in order to allow the commitment of EDF to invest in this nuclear project and serve a public interest objective as long as they do not have distortive effects on competition and trade contrary to the common interest, (vii) the measures in question do not constitute a public contract or a concession, but a simple subsidy, since they do not allow the UK to require EDF either to build Hinkley Point C or to supply electricity.

In October 2015, EDF and the UK Government announced updates to the October 2013 provisional agreement of commercial terms of the deal for the £16 billion cost of construction of Hinkley Point C. The estimated price of construction has since risen, and EDF said that the increase announced in 2017 resulted mainly "from a better understanding of the design adapted to the requirements of the British regulators, the volume and sequencing of work on site and the gradual implementation of supplier contracts". EDF maintained the official construction-start target-date as "mid-2019" and the "initial delivery objective at the end of 2025". However, EDF admitted that the beginning of construction for Hinkley Point C in mid-2019 can only happen if "the final design, which is on a tight schedule, is completed by the end of 2018". The Hinkley Point C project is on-going but still not officially under construction, illustrating the case where a significant share of investment and work can be realized before even entering the official construction phase. There was no arbitration case in the "pre-construction phase" of Hinkley Point C as in the case of Olkiluoto in Finland, however, a question that arises is whether a specific construction timetable can be achieved on time and be more likely to meet the start-up date that will be officially announced.

3 European Court of Justice's decision on environmental impact of lifetime prolongation of nuclear power plants – The case of Belgium and the Netherlands

In the context of investment plans for nuclear plant life-extension for energy security reasons, it is interesting to mention the case of Belgium and the operational life prolongation of two nuclear reactors, Doel 1 and Doel 2, which are on a site close to the Dutch border.[19] In 2003, the national legislation of Belgium anticipated the phase-out of nuclear energy for 2015 limiting the lifetime of nuclear power plants to 40 years. However, in 2015, a modification of the above legislation was decided by the Belgian nuclear regulator for the extension of the lifetime of Doel 1 and Doel 2 for another 10 years putting on hold the plans for their decommissioning. A specific action plan aiming at the continuation of the operation of these reactors beyond their design life and the modernization of aging facilities ensuring the compliance with up-to-date safety measures was submitted by Electrabel, which is the owner and operator of the

---

19 "EU court warns Belgium over nuclear reactors", *EURACTIV*, 29 July 2019, https://www.euractiv.com/section/electricity/news/eu-court-warns-belgium-over-nuclear-reactors/; "EU court rules on extended operation of Doel units", *World Nuclear News*, 29 July 2019, https://www.world-nuclear-news.org/Articles/EU-court-rules-on-extended-operation-of-Doel-units; "Belgium Loses Court Battle Over Nuclear Plant Extensions", *Courthouse News Service*, 29 July 2019, https://www.courthousenews.com/belgium-loses-court-battle-over-nuclear-plant-extensions/; "CJEU confirms that environmental impacts of life-time extensions must be assessed", *ClientEarth*, 24 October 2019, https://www.clientearth.org/cjeu-confirms-that-environmental-impacts-of-life-time-extensions-must-be-assessed/ (websites accessed 4 June 2020).

nuclear power stations, and approved by the Belgian regulator and the Federal Agency for Nuclear Control (FANC) during that period. However, two Belgian environmental associations – Inter-Environnement Wallonie and Bond Beter Leefmilieu Vlaanderen – brought a case before Belgium's Constitutional Court for the annulment of the law, saying the operating life prolongation had been adopted without carrying out environmental impact studies and without public consultation.

The case was referred to the European Court of Justice. Three were the main issues addressed by the EU Court: (i) possible infringements of the Environmental Impact Assessment (EIA) Directive, (ii) possible infringements of the Habitats Directive,[20] (iii) keeping valid the contested Belgian law if the Directives had indeed been breached. In July 2019, the European Court of Justice ruled that Belgium should have realized an environmental assessment before the adoption of the extension because the proposed project was "of a comparable scale, in terms of risks of environmental impact, to the initial commissioning of those power stations" taking into account the length of the extension and the nature of the necessary works. The judgment of EU Court thus ensured that a broad assessment as to whether the impact of the extension requires EIA is undertaken. Particularly, the primary advisor to the EU Court, advocate general Juliane Kokott came to the conclusion that the Espoo Convention,[21] and the EIA Directive (85/337/EEC) in conjunction with the Aarhus Convention,[22] oblige Belgium to carry out an EIA for the life-time extension of the Doel 1 and 2 nuclear reactors. According to the ruling, an exemption from the environmental impact study can be obtained only if the national court "establishes that the risk to the security of electricity supply is reasonably likely, and that the project is as urgent in nature as to justify the lack of such an assessment".

---

20 The Habitats Directive (Council Directive 92/43/EEC of 21 May 1992) ensures the conservation of a wide range of rare, threatened or endemic animal and plant species. Some 200 rare and characteristic habitat types are also targeted for conservation in their own rights.

21 The Espoo (EIA) Convention sets out the obligations of Parties to assess the environmental impact of certain activities at an early stage of planning. It also lays down the general obligation of States to notify and consult each other on all major projects under consideration that are likely to have a significant adverse environmental impact across boundaries.

22 Under the obligations of the Aarhus Convention, any decision to prolong the lifetime of nuclear power reactors should be preceded by meaningful public participation concerning environmental issues.

The EU Court said that no such situation appeared to be met in this case and the works indicated in the action plan affected the physical reality of the sites concerned. Concerning the Habitats Directive, the EU Court argued that, similarly to the EIA Directive, an appropriate assessment had to be carried out prior to the adoption of the law prolonging the lifetime and the court in Belgium must verify if "the need to prevent a genuine and serious threat of interruption to the supply of electricity" overrides the risk that the project would affect a site hosting a priority natural habitat type or a priority species. Finally, the EU Court decided that the Belgian law as it was modified in 2015 could be kept in place until the environmental infringements have been remedied. Hence, the EU Court concluded that the necessary assessment studies could still be conducted to retroactively regularize the project while the action plan is being executed or even after it has been completed, and, as the Doel 1 and Doel 2 reactors are on a site close to the Dutch border, the Netherlands must be involved in a "transboundary assessment procedure".

Alexandros Chatzinerantzis[*]

13 Climate-change related disputes – Overview and trends

Abstract: Climate change is increasingly occupying the courts. The number of claims from organizations, citizens, contractors, governments and even investors and shareholders against global entities and/or entire states is constantly rising. Essentially, these disputes can be divided into four main categories, all of which in their own way have serious implications not only for the policies of states but also for the business models and the corporate policies of companies. These categories include Climate Enforcement Disputes, Commercial Litigation and Arbitration, Investor-State Arbitration and Investor and D&O claims.

Keywords: Climate Change Litigation, International Arbitration, Investor-State Arbitration (ICSID), Energy Charter Treaty (ECT), Environmental Disclosure Duties (ESG-Package), Shareholder- and D&O-claims

Introduction

This chapter, based on a presentation held at the 4th Athens Conference on European Energy Law on 27 September 2019, shall give an overview of key trends in climate-change-related disputes and what we may be facing in the future. The discussion about climate change is omnipresent. Expectedly, for most, certain headlines or pictures come to mind when the term "climate change" is heard: We are faced with more destructive Category 5 hurricanes that are developing, massive wildfires are blazing on every continent. Polar ice and glaciers are melting at an unprecedented rate, leading to an accelerated sea-level rise, while oceans are becoming warmer and more acidic. Worsening droughts and biblical floods threaten to render large areas of the Middle

---

[*] The author is partner at the international law firm Linklaters LLP, Frankfurt a.M. Alexandros heads the German Dispute Resolution Group and is a renowned expert on disputes in the energy sector.

East, the Persian Gulf, North Africa and South Asia uninhabitable.[1] In a nutshell: Climate change does affect every human being – and is also reaching the courts.

Range of climate-related disputes

Climate-related disputes are not limited to the question of the legal responsibility of companies for the global consequences of climate change.[2] The range of observed climate change-related disputes is vast and covers numerous legal issues across multiple fields of law and various causes of action with a wide range of claimants and defendants from multiple sectors. This chapter takes up this comprehensive field of climate change-related disputes and offers a survey of the essential key areas. First, the rapidly emerging area of Climate Enforcement Disputes is introduced, before two sub-categories – claims to accelerate policy changes and claims for compensation – are presented in more detail. Secondly, the article shifts its focus to the more conventional field of Commercial Litigation and Arbitration and thereupon discusses Investor-State Arbitrations as another important type of a climate-drawn dispute. The article concludes with the introduction of shareholder and D&O claims as the fourth main pillar of climate-related disputes.

1 Climate Enforcement Disputes

According to the database of the Sabin Centre for Climate Change at the University of Columbia in New York, more than 1,400 climate change-related cases have been brought before courts so far in at least 28 countries all around the world.[3] As a general trend, climate change litigation continues to see a geographic expansion. There are now cases in the Americas, Asia, Europe and the Pacific Region, which either aim to influence public and private accountability to instigate broader policy debates and change, or – on a less visible

---

1 Al Gore, "The Climate Crisis Is the Battle of Our Time, and We Can Win", *New York Times*, 20 September 2019, https://www.nytimes.com/2019/09/20/opinion/al-gore-climate-change.html.
2 Alexandros Chatzinerantzis and Markus Appel, "Climate Change Litigation – The Liability of CO_2-Emitters under German Law", *Carbon & Climate Law Review* (2019), 280–286; Alexandros Chatzinerantzis and Markus Appel, "Haftung für den Klimawandel", *Neue Juristische Wochenschrift* (2019), 881–886.
3 "Climate Change Litigation Database", *Columbia Law School and Arnold & Porter*, http://climatecasechart.com/about/.

scale – simply expose courts to climate change arguments which, until recently, did not appear in a legal context. However, in both cases, the procedures can cause considerable impact on climate policy enforcement by governments or private parties.[4] Climate change disputes hence either deliberately aim at or inadvertently result in enforcing climate change action.

1.1 Cases brought forward against government agencies to change climate policy and conduct

The first category includes claims pursued by NGOs or other activists against governments and – most recently – also private companies to enforce existing climate policies or to accelerate policy changes. The most prominent example in this regard is the so-called *Urgenda* case "Urgenda Foundation v. State of the Netherlands". A Dutch environmental group, the *Urgenda* Foundation, and 900 Dutch citizens sued the Dutch government to require it to do more to prevent global climate change. The claimants brought forward the Dutch climate targets,[5] violated provisions of the Dutch Constitution, the European Convention on Human Rights and the government's duty of care under the Dutch Civil Code.[6] And they succeeded: The court in the Hague ordered the Dutch state to limit greenhouse gas emissions ("GHG") to 25 % below 1990 levels by 2020, finding that the state has a duty to take climate change mitigation measures due to the "severity of the consequences of climate change and the great risk of climate change occurring". The Hague Court of Appeal upheld this decision on 9 October 2018.[7] This case has been the blueprint for several copycat proceedings, inter alia the so-called *People's Climate Case*,[8] which was brought forward by several individuals against the European Parliament and

---

4 Kim Bouwer, "The unsexy future of climate change litigation", *Journal of Environmental Law* (2018), 484.
5 Until 2011, the Dutch government maintained a reduction target of GHG emissions of 30 % for 2020. However, in order to align with the EU's guidelines, the State later adjusted this mark to 20 %; see "Hague Court of Appeal Requires Dutch Government to meet greenhouse gas emissions reduction by 2020", *Harvard Law Review* 132 (2019), 2090, https://harvardlawreview.org/2019/05/state-of-the-netherlands-v-urgenda-foundation/.
6 Ibid., p. 2091.
7 See Court of Appeal in The Hague, 9 October 2018, AB 2018, 417 m.nt. GA van der Veen, Ch. W. Backes, *Staat der Nederlanden v. Stichting Urgenda*, 3.9, rec. 76.
8 General Court, order of 8 May 2019, case T-330/18, ECLI:EU:T:2019:324, *Carvalho and Others v. Parliament and Council*.

the Council of the European Union. The claimants sought to compel the EU to take more stringent GHG emission reductions and alleged that the EU's existing target to reduce domestic GHG emissions by 40 % by 2030, as compared to 1990 levels, is insufficient to avoid dangerous climate change and threatens claimants' fundamental rights of life, health, occupation, and property. This claim was dismissed on procedural grounds. An appeal is currently pending before the European Court of Justice.[9] Another prominent example is the claim of three German families and Greenpeace Germany against the German federal government known as the *German Climate Case*.[10] The claimants are organic farmers asserting they are already experiencing the impacts of climate change. The claimants cited the *Urgenda* decision and argued that the German government failed to meet its GHG emission reduction targets and thus violated their constitutional rights to life and health, property and occupational freedom. However, it was not really convincing to refer to the *Urgenda* case, which ordered the Dutch government to reduce its GHG emissions to 25 % below 1990 levels by 2020, whereas Germany has already reduced its emission by 32 %. Hence on 31 October 2019, the Administrative Court of Berlin dismissed the claim.

Comparable cases have surfaced in France, where in December 2018, four non-profit organizations initiated legal proceedings against the French government for inadequate action on climate change.[11] A further claim has been brought forward against France in early 2019.[12] The common characteristic in these cases is that they were directed against governments or supranational institutions. However, as the recently filed case of *The People v. Shell* shows,[13] the effort of climate change litigation has shifted and is now also directed against

---

9 European Court of Justice, appeal brought on 23 July 2019, case C-565/19 P, by Armando Carvalho and Others against the order of the General Court delivered on 8 May 2019 in case T-330/18, *Carvalho and Others v. Parliament and Council*.

10 Administrative Court of Berlin, judgment of 31 October 2019, case 10 K 412.18.

11 Administrative Court of Paris, 14 March 2019, *Notre Affaire à Tous and others v. France*, http://blogs2.law.columbia.edu/climate-change-litigation/wp-content/uploads/sites/16/non-us-case-documents/2019/20190314_NA_press-release.pdf.

12 Conseil d'Etat, lawsuit brought on 23 January 2019, *Commsune de Grande-Synthe v. France*, http://climatecasechart.com/non-us-principle-law/european-convention-on-human-rights/.

13 District Court of The Hague, lawsuit brought on 5 April 2019, *Milieudefensie et al. v. Royal Dutch Shell plc.*, http://climatecasechart.com/non-us-case/milieudefensie-et-al-v-royal-dutch-shell-plc/.

the private sector itself: In this case the environmental group *Milieudefensie/ Friends* together with more than 17.000 citizens as co-claimants demand from *Shell* a modification of its corporate policy and business model and seek a ruling from the Dutch court that *Shell* must reduce its GHG emissions by 45 % by 2030 compared to 2010 levels and to zero by 2050, in line with the Paris climate agreement. This case also builds on the *Urgenda* decision. However, claimants try to extend this decision to private companies by arguing that *Shell* has, for a long time, had knowledge of the dangers of climate change. Given the Paris agreements' goals and the (alleged) scientific evidence for the reasons of climate change, *Shell's* misleading statements in this regard, and inadequate measures to reduce GHG emissions constitute hazardous negligence. Another notable case has been brought forward in Poland. The environmental law organization *ClientEarth* filed a lawsuit against the operator of Europe's largest power plant and Europe's greatest source of GHG pollution, *Belchatow, Polska Grupa Energetyczna*.[14] The lawsuit seeks to block the plant operators from burning lignite – or requires measures to reduce carbon dioxide emissions – by 2035. The claimant based its claim on Article 323 of Poland's 2001 Environmental Protection Law, which permits ecological organizations to demand that preventive measures be taken when an activity harms the environment as a common good. Just recently, six NGOs have sued energy the company *Total* in France over an oil project in Uganda and Tanzania.[15] The claimants allege that *Total* failed to adequately assess the project's threats to human rights and the environment as a result of their business practices. The court in Paris scheduled a hearing on 8 January 2020.

1.2 Cases brought forward to seek damages for financial redress

The second category of cases comprises civil claims pursued by private parties or municipalities against high-emitting companies, so-called carbon majors, seeking damages or financial redress for their alleged responsibility for climate change. The main argument in these cases is that the emissions of a small group of corporations over time had significantly contributed to climate change. One of the most prominent examples in this regard – which also illustrates the truly

---

14 "Poland: The New Battleground For Climate Litigation", *Energy and Climate Law blogspot*, 7 October 2019, http://energyandclimatelaw.blogspot.com/2019/10/poland-new-battleground-for-climate.html/.
15 Nanterre High Court, complaint brought on 23 October 2019, *Les Amis de la Terre v. Total*, http://climatecasechart.com/non-us-case/friends-of-the-earth-et-al-v-total/.

global nature of climate change disputes – is the claim of the Peruvian farmer *Saul Luciano Lliuya* against *RWE*, Germany's largest electricity producer, for its emissions in the past 250 years which allegedly represent an estimated 0.47 % of the world-wide GHG emissions caused in this period. Thus, contributing to global warming, which in turn would be responsible for the melting of the Palcacocha glacier near the plaintiff's property. The claimant sought the costs of securing his property in the Andean town of Huaraz, a potentially flood-prone area about 450 km north of Lima, from potential floods caused by the melting glacier. Although the regional court in Germany initially dismissed this claim,[16] the appeal is currently pending before the Higher Regional Court of Hamm which ordered the taking of evidence by an expert.[17] Another variation of civil claims against corporations is based on the allegation that corporations continue to produce fossil fuels while knowingly concealing the climate risks. Vivid examples are the claims brought forward by the *State of Rhode Island*,[18] or the cities of *New York*,[19] *San Francisco*,[20] and *Oakland*,[21] against several petrol companies like *Chevron, BP* or *ExxonMobil*. These lawsuits seek to hold fossil fuel companies liable for causing climate change impacts that adversely affect the residents, businesses, state-owned facilities, real property and other assets, notably due to the cities' and state's vast coastline. The alleged harms include a rise in sea levels, increased frequency and severity of flooding, extreme precipitation events, as well as ocean warming and acidification. The claimants

---

16 Regional Court of Essen, judgment of 15 December 2016, case 2 O 285/15, *Neue Zeitschrift für Verwaltungsrecht* (2017), 734–736.
17 Higher Regional Court of Hamm, decision of 30 November 2017, case I-5 U 15/17, *Zeitschrift für Umweltrecht* (2018), 118–119.
18 United States District Court for the District of Rhode Island, 2018, case 1:18-cv-00395, *State of Rhode Island v. Chevron Corp et al.*; United States Court of Appeals for the First Circuit, 2019, case 19-1818, *State of Rhode Island v. Shell Oil Products Co. et al.*; Supreme Court, 2019, case 19A391, *BP p.l.c. et al. v. State of Rhode Island*, http://climatecasechart.com/case/rhode-island-v-chevron-corp/.
19 United States District Court for the Southern District of New York, 2018, case 1:18-cv-00182, *City of New York v. BP p.l.c. et al.*, http://climatecasechart.com/case/city-new-york-v-bp-plc/.
20 Texas Courts of Appeals, 2018, case 02-18-00106-CV, *City of San Francisco v. Exxon Mobil Corp*, http://climatecasechart.com/case/re-exxon-mobil-corp/.
21 United States Court of Appeals for the Ninth Circuit, 2018, case 18-16663, *City of Oakland et.al. v. BP p.l.c. et al.*, http://climatecasechart.com/case/people-state-california-v-bp-plc-oakland/.

2 Commercial Litigation and Arbitration

The next type of climate-related disputes can be classified as traditional commercial Litigation and Arbitration. We are currently in the middle of a tremendous transition in the energy sector worldwide. One of the most challenging issues is the question of decarbonisation. In this regard, we seem to be faced with a dilemma. On the one hand, experts estimate that worldwide energy consumption will rise by around 30 % by 2040, which is closely related to the constant rise of the living standards in large parts of the world. Currently, 80 % of the world's population lives in counties where more energy demand will go hand in hand with significant improvements in the quality of life (electric light, heating, kitchen devices, etc.). This growth of energy demand automatically leads to further growth of carbon dioxide emissions worldwide. However, to meet the Paris climate goals, a reduction of carbon dioxide emissions by 45 % is needed. Thus, to get anywhere close to a path which is consistent with the Paris climate goals will require a speed of change and transition in the global energy system, which is truly unprecedented. Achieving this pace of transition will require a comprehensive set of policy measures, inter alia in the energy, transport, and industrial sectors. One major measure, for example, would be the total phase-out of coal-fired power generation which is currently discussed in various countries such as the UK, the Netherlands, Germany, France and Greece. The decommissioning of so many coal-fired power plants[22] is not going to happen silently. Such a disruptive change will, without doubt, lead to numerous frictions and tensions along the value chains interwoven with the power plants (jobs, coal logistics, suppliers, plant and system construction, maintenance, purification, etc.).

Such tensions will automatically lead to numerous disputes, inter alia in connection with "stranded assets".[23] Further, the retired capacities have to be replaced. Hence, we are going to see a huge number of construction activities all over the world with onshore and offshore wind parks, large-scale PV farms, LNG terminals, substantial expansions of the gas and the energy grids,

---

22 In Germany alone, approx. 43 GW of coal power plants will be decommissioned.
23 Sini Matikainen, "What are stranded assets?", *Grantham Research Institute on Climate Change and the Environment*, 23 January 2018, http://www.lse.ac.uk/GranthamInstitute/faqs/what-are-stranded-assets/.

interconnectors and so on. With developers, contractors and investors globally involved on such an ambitious scale, the scope for commercial disputes is vast and will inevitably arise. These disputes will likely consist of, inter alia, EPC contractor disputes about damages, defects or delays, late commissioning, post-M&A disputes and any kind of disputes with suppliers or other involved third parties. Another field of disputes relates to price-adjustment claims. As we are facing substantial changes on the energy markets with multiple effects on all market participants, we are likely to see claims based on hardship, or *force majeure* claims, where risks have not been anticipated by the parties. Whenever new risks materialise, parties regularly seek to mitigate and allocate these risks between themselves contractually. Unsurprisingly, many contracts now include obligations to comply with environmental and human rights or sustainability obligations, and commitments to put in place back-to-back arrangements with counterparties further down the value chain. Disputes over those obligations will eventually arise. Arbitration institutions have also identified the huge potential for climate-related disputes. The ICC has seen a steady increase in its caseload in the energy sector, with the registration of 155 new energy cases, that represented 19 % of its caseload in 2017. It has thus set up a task force on arbitration and climate change-related disputes.[24] "The Task Force predicts that climate change-related disputes will increase exponentially in the coming years and suggests that the ICC is best positioned to deal with such disputes, given its active role in negotiating on issues of climate change on behalf of businesses and industry. It also has procedures in place to allow for unique solutions to address climate change-related issues, including recourse to arbitrators and experts with appropriate scientific and climate change-related expertise."[25]

3 Investor-State Arbitration

Moreover, since there is substantial regulation to be put into place, multiple disputes with regulatory authorities are looming. One example is the Nord Stream 2 pipeline, where a Gazprom subsidiary has applied to the European Court of Justice to annul the amended Gas directive. The applicants challenge

---

24 Pierre Duprey and Roland Ziadé, "Arbitration and meeting the demands of the future energy sector", *Guest Blog ICC*, 1 April 2019, https://iccwbo.org/media-wall/news-speeches/arbitration-and-meeting-the-demands-of-the-future-energy-sector/.
25 Cherine Foty, "France: Resolving Climate Change Disputes Through Arbitration", *Mondaq*, 16 September 2019, https://www.mondaq.com/france/Environment/845190/Resolving-Climate-Change-Disputes-Through-Arbitration.

the alleged discriminatory extension of EU internal market rules, applicable to gas pipelines importing from non-EU countries. Further, on 26 September 2019, the Nord Stream 2 investors served a notice of arbitration against the EU under the Energy Charter Treaty (ECT).[26] This leads us to the next type of disputes: Investor-State Arbitrations, inter alia disputes under the ECT or under bilateral investment treaties. As pointed out, significant investments will be required to achieve the Paris climate goals. In 2017, the OECD estimated that annual investments of USD 6.3 trillion are needed until 2030. A substantial portion of this will come from private investments, including foreign direct investment (FDI). Pursuant to estimations from EU-institutions, private funding of EUR 175 to 290 billion per year is needed alone in the EU.[27] With such an increase in new FDI, there will automatically be an increase in disputes between investors and host states. Furthermore, disputes will arise in the context of pre-existing investments. As described, to achieve the pace of transition needed to meet the Paris climate goals, it will require a comprehensive set of policy measures. However, we all know that fundamental changes to the investment environment often lead to investment disputes.

A prime example is the large number of over 40 claims against Spain under the ECT following several reforms of the renewable energy system between 2012 and 2014. Some of these cases have already been decided – mostly in favour of the claimants – stating, in general terms, that regulatory regimes cannot be radically altered so as to deprive investors of the value of their investment in an unpredictable way.[28] A further recent example is the fate of *Uniper's* relatively new coal-fired power plant *Maasvlakte 3* in the Netherlands: this power plant, which became operative in 2016 after a year-long construction phase that cost

---

26 Anca Gurzu, "Nord Stream 2 sues the EU over new gas rules", *POLITICO*, 26 September 2019, https://www.politico.eu/article/nord-stream-2-sues-the-eu-over-new-gas-rules/.

27 "ESG-Initiative der EU: Die Ziele sind Hoch Ambitioniert", *Der Assetmanager*, 30 April 2019, https://www.derassetmanager.de/esg-initiative-der-eu-die-ziele-sind-hoch-ambitioniert/.

28 ICSID, award of 4 May 2017, case ARB/13/36, *Eiser Infrastructure Limited and Energia Solar Luxembourg S.à r.l. v. The Kingdom of Spain*. The tribunal underlined that the state has the right to regulate and to modify regulatory regimes according to evolving circumstances, and that the FET standard would not lead to a right of investors to regulatory stability. However, it found that the ECT protected Eiser "against total and unreasonable change that they experienced here", para. 363. For a case decided in favour of State see SCC, award of 21 January 2016, case 062/2012, *Charanne B.V. and Construction Investments S.a.r.l. v. Spain*.

the company EUR 1.6 billion, is affected by the plans of the Dutch government to decommission all coal-fired power plants by 2030 the latest.[29] Another recent case is the claim of Austrian engineering company *Strabag* against Germany in the context of the reform of the German renewables feed-in system in 2016. On 20 September 2019, *Strabag* and two affiliates, *Erste Nordsee-Offshore Holding* and *Zweite Nordsee-Offshore Holding* filed an ECT complaint about changes of the Renewable Energy Sources Act (EEG) that affected its plans to develop wind projects in the North Sea.[30]

4 Investor and D&O claims

Last but not least, we see a significant rise in regulatory investigations of corporates over disclosures of material climate-related risk, and increasing investor activism, including investor claims. Claimants in several jurisdictions have made claims against investment funds and companies for failing to incorporate climate risks into their decision-making, and for failing to disclose or negligently misstating climate risks to their beneficiaries. For example, *ExxonMobil* has been sued in New York by a shareholder alleging that *Exxon* misled investors, by understating how much risk climate change itself and regulations addressing the climate crisis pose to the company's assets.[31] The Attorney General of the State of New York essentially accused *ExxonMobil* of fraudulently misleading shareholders by using two separate sets of metrics to calculate financial risks relating to potential future costs with regard to governmental steps to cut GHG emissions: One set was shared with investors and was based on a higher estimated cost of carbon (suggesting stricter limits on emissions had been considered). Another set was used internally for decision-making on investments using a lower figure and reflecting more lenient regulations. The alleged impact of such a practice is that internally high-polluting projects appeared as more attractive investments. The claimant alleges that *Exxon* misled investors into thinking that the company was better placed to manage

---

29 Bart H. Meijer, "Netherlands to ban coal-fired power plants in blow to RWE", *Reuters*, 18 May 2018, https://www.reuters.com/article/us-netherlands-energy-coal/netherlands-to-ban-coal-fired-power-plants-in-blow-to-rwe-idUSKCN1IJ1PI.
30 Cf. ICSID, 2019, case ARB/19/29, *Strabag SE, Erste Nordsee-Offshore Holding GmbH and Zweite Nordsee-Offshore Holding GmbH v. Federal Republic of Germany*.
31 New York Supreme Court, 2018, case 452044/2018, *People of the State of New York v. ExxonMobil Corp.*, http://climatecasechart.com/case/people-v-exxon-mobil-corporation/.

climate-related risks than it was in practice and seeks compensation between USD 476 million and USD 1.6 billion.

The trial began on 22 October 2019. In relation to negligent misstatement claims, the key argument made by claimants focuses on the company having misstated the potential impacts of climate change on its business or the ability of the company to manage these risks. Similar claims could be brought by shareholders or investors, who invested in high GHG emitting companies, based on publicly-made statements or reports on the companies' exposure profiles. Where there is then fall in share price because of climate-related developments, it is potentially open to those investors to bring a claim in respect of the loss of value of their shares. In addition to litigation for failing to incorporate climate risk into investment decisions, several cases have sought to require improved disclosure of climate risks to investors and shareholders. For instance, in *Abrahams v. Commonwealth Bank of Australia*,[32] shareholders alleged that the bank violated the Corporations Act of 2001 with the issuance of its 2016 annual report, which failed to disclose climate change-related business risks with regard to an imminent investment in the controversial Carmichael coal mine.[33] While disclosure does not per se insulate a firm from climate-related litigation, failing to report climate risks and/or comply with recommendations is likely to increase litigation risks. In this regard, one of the most important recent developments is the environmental, social and governance (ESG) package of the European Commission presented in May 2018 and its implication in the energy sector.[34] This so-called ESG-proposal is part of a broader initiative on sustainable development and aims at transforming Europe's economy into a greener, more resilient and circular system. ESG factors encompass a wide range and include, inter alia, climate change/carbon emissions, resource management, pollution and waste, environmental opportunities (e.g. clean tech) as environmental factors.[35]

---

32 Cf. Federal Court of Australia, 2017, case VID879/2017, *Abrahams v. Commonwealth Bank of Australia*.
33 Further examples are Massachusetts Superior Court, 2019, case 19-3333, *Commonwealth v. Exxon Mobil Corp.*; District Court Southern District of Texas, 2016, case 4:16-cv-03484, *Fentress v. Exxon Mobil Corp.*; District Court Northern District of Texas, 2016, case 3:16-cv-3111, *Ramirez v. Exxon Mobil Corp.*
34 For an overview see the Commission's Action Plan: Financial Sustainable Growth, COM(2018) 97 final of 8 March 2018.
35 Cf. the Proposal for a Regulation of the European Parliament and of the Council amending Regulation (EU) 2016/1011 on low carbon benchmarks and positive carbon impact benchmarks, COM(2018) 355 final; labour management, health and safety, product safety, supply chain as social factors on the social side, cf. Proposal for a

The proposed regulation introduces additional requirements to existing elements of the IORP II-Directive 2016/2341, including a dedicated and coherent disclosure framework on the integration of ESG risks. This must be used by financial market participants both in investment decisions or advisory processes and has to be made transparent to end-investors. It is obvious that this will put financial market participants in a tight spot: Not only will they have to disclose sustainability risks and comply with the transparency provisions, but investors most likely will demand ESG-friendly investments as well. This inevitably will be the gateway for legal disputes: NGOs will sue if transparency and disclosure obligations are not met, investors will sue if ESG risks are whitewashed. Ultimately, even the executive bodies could be held liable for any infringement of ESG considerations. As a matter of fact, incorrect financial communications and violations of laws or regulatory provisions are still the most common ground for actions against directors and officers. Regulatory enforcement has also increased over the past years and led to more litigation against company executives (D&O-litigation).

Conclusions

Climate change-related claims have already started to materialize in various shapes and sizes. It is expected that the number of those disputes will increase significantly. Over the last years, top managers had identified reputational loss and data theft as the biggest risks for their companies. Nowadays, climate change is at the top of the risk list: Three-quarters of CEOs see it as a key challenge to tailor their business models to greater environmental compatibility, less carbon dioxide emissions and more clean technology.[36] Put plainly: climate-related dispute risk is now a part of corporate reality.

Regulation of the European Parliament and of the Council on the establishment of a framework to facilitate sustainable investment, COM(2018) 353 final; board diversity, executive pay, business ethics, corruption and instability, tax transparency and CSR as governance factors, cf. Proposal for a Regulation of the European Parliament and of the council on disclosures relating to sustainable investments and sustainability risks and amending Directive (EU) 2016/2341, COM(2018) 354 final.

36 "Agile or Irrelevant – Redefining resilience – Global CEO Outlook 2019", *KPMG*, 2019.

Prof. Dr. Matthias Knauff, LL.M. Eur.*

14 Representative actions in the energy sector

Abstract: Representative actions are of high relevance to the energy sector in Germany. Recognized environmental protection associations are entitled to take legal action against environmentally relevant energy industry projects nearly without any restrictions according to the Law on supplementary provisions on remedies in environmental matters (Umweltrechtsbehelfsgesetz – UmwRG). As a result of the qualification of previously existing rules of preclusion in German law as contrary to European law by the ECJ, the requirements for a successful representative action have become even lower. With the upcoming change of the directive on representative actions for the protection of the collective interests of consumers, representative actions will get even higher importance in the energy sector.

Keywords: representative action, Aarhus Convention, environmental associations, preclusion, consumer protection

Introductory remarks

In the course of the energy transition, it is well known that there is no way around the establishment of new generation plants for "green power" as well as the extension of the power grids. At the same time, however, in the German population, there is considerable resistance to new energy infrastructure. This is reflected in a protest and lawsuit wave. Although the energy transition as such is also, essentially, a project of environmental and climate protection, it is increasingly coming into conflict with aspects of nature and landscape conservation. As a result, individual citizens and citizens' initiatives take action against specific projects in addition to environmental associations. At the same time, the latter have become more professionalized throughout the past few years and are increasingly effectively fulfilling their function as guardians and preservers of the environment anchored in society, even under recourse to legal protection. The following chapter presents the current form of representative action in German environmental law, which has considerable significance for the energy sector. In the future, it will also be largely covered by the directive

---

* Prof. Dr. Matthias Knauff, LL.M. Eur. is holder of the Chair of Public Law, in particular Public Commercial Law at the Friedrich-Schiller-University of Jena.

on representative actions for the protection of the collective interests of consumers, which will create new challenges.

1 Representative actions regarding environmental matters

With regard to environmental matters, representative actions have received widespread recognition in German law. At the same time, they are part of the 'general risks' to which the energy industry is regularly exposed due to the environmental relevance of its projects.

1.1 Development

The traditional claim system, which was based on subjective law, was alien to the representative action. Only the Aarhus Convention (1998)[1] initiated a change in this regard. Nevertheless, the German legislature remained hesitant, and the triumphant progress of the environmental law representative action was driven substantially by the legal development on European level. Of particular importance in this regard was Directive 2003/35/EC of the European Parliament and of the Council of 26 May 2003 providing for public participation in respect of the drawing up of specific plans and programmes relating to the environment and amending with regard to public participation and access to justice Council Directives 85/337/EEC and 96/61/EC,[2] which was intended to transpose the Aarhus Convention into EU law. Following the replacement of Directive 96/61/EC in the meantime, the relevant provisions are now contained in Directive 2010/75/EU of the European Parliament and of the Council of 24 November 2010 on industrial emissions (integrated pollution prevention and control).[3] For the first time, an environmental association complaint was introduced into German law by § 61 of the Federal Nature Conservation Act (Bundesnaturschutzgesetz – BNatSchG, 2002).[4] It was not until 2006 – and thus one year before the ratification of the Aarhus Convention by the Federal Republic of Germany – that the law on supplementary provisions on remedies in environmental matters pursuant to EU Directive 2003/35/EC

---

1   UNECE Convention on Access to Information, Public Participation in Decision-making and Access to Justice in Environmental Matters (Aarhus Convention), BGBl. II 2006, 1251, II 2007, 1392.
2   OJ 2003 L 156, 25 June 2003, 17.
3   OJ 2010 L 334, 17 December 2010, 17.
4   Gesetz über Naturschutz und Landschaftspflege, BGBl. I 2002, 1193.

(Umweltrechtsbehelfsgesetz – UmwRG),[5] was issued. Political efforts were made to keep the representative action lawsuit 'small' and to ensure compatibility with the existing legal protection system,[6] in particular restricting the right of action of environmental associations to cases where individuals are also entitled to, which in 2011 was classified by the European Court of Justice as being contrary to European law.[7] A separation of the environmental association's legal action from the requirement of a subjective infringement then ensued with the amendment of the UmwRG in 2013, which finally established the altruistic collective action on environmental matters without any restrictions.[8]

1.2 Design

The UmwRG provides for the possibility of representative action for numerous environmentally relevant decisions. Projects of the energy industry are affected in many cases. The admissibility of the environmental action of the association is defined in § 2 (1–3) UmwRG. According to this, domestic and foreign environmental associations have a comprehensive right of action within the scope of their respective statutory duties with regard to possible environmental law violations through administrative decisions. The prerequisite for this is their recognition according to § 3 UmwRG, which requires a request of the association directed to this. The standardized conditions for this are not very demanding,[9] and are in the end contrary only to the right to sue spontaneously by local citizens' initiatives arising due to concrete concerns.[10] It is necessary that the association "1. according to its statutes promotes ideological and not only temporary predominantly the objectives of environmental protection, 2. has existed for at least three years at the time of recognition and has been active in this period within the meaning of point 1, 3. provides the guarantee for proper performance of the task, in particular for proper participation in

---

5   Environmental Legal Remedy Act (Umweltrechtsbehelfsgesetz), BGBl. I 2006, 2816.
6   Cf. BT-Drucks. 16/2495, p. 12.
7   European Court of Justice, judgment of 12 May 2011, case C-115/09, ECLI:EU:C:2011:289, *Bund für Umwelt und Naturschutz Deutschland, Landesverband Nordrhein-Westfalen eV v. Bezirksregierung Arnsberg*.
8   BT-Drucks. 17/10957, pp. 15 et seq.
9   In detail, Sabine Schlacke, in: *Verwaltungsgerichtsordnung mit Nebengesetzen*, ed. Klaus Ferdinand Gärditz (Köln: Carl Heymanns Verlag, 2018), para. 3 UmwRG rec. 12 et seq.
10   Cf. Franziska Heß, "Aktivierung der Umweltverbandsklage", *Zeitschrift für Umweltrecht* (2018), 686, 687.

official decision-making procedures; the nature and extent of their previous activities, the circle of members and the efficiency of the association must be taken into account, 4. pursue charitable purposes within the meaning of § 52 of the Tax Code [Abgabenordnung] and 5. allowing each person to join as a member who supports the goals of the association; Members are persons who receive full voting rights in the general meeting of the association upon entry; in the case of associations whose membership consists of at least three-quarters of legal persons, the condition in the first sentence of Part 1 may be waived, provided that the majority of these legal persons fulfill this condition". According to § 2 expressly not required for the admissibility of the action is the assertion of violation of the association's own rights. In that regard, the provision deviates significantly from the "normal case" of § 42 (2) Code of Administrative Court Procedure (Verwaltungsgerichtsordnung – VwGO), according to which an action before the Administrative Tribunal,[11] is regularly admissible only "if the plaintiff claims that his/her rights have been violated by the administrative act or its refusal or omission". As a result, the association's claim under the UmwRG – despite individual parallel regulations in the areas of consumer protection,[12] equality of disabled people,[13] and animal welfare (in most federal states),[14] – continues to qualify as a foreign body in the German legal protection system.

---

11 Regarding the litigation power in civil procedure cf. German Federal Court of Justice, judgment of 25 November 2004, case I ZR 145/02, *Neue Juristische Wochenschrift* (2005), 1656, 1657.

12 Para. 606 sec. 1 Zivilprozessordnung (ZPO); para. 3 sec. 1 sentence 1 Gesetz über Unterlassungsklagen bei Verbraucherrechts- und anderen Verstößen (Unterlassungsklagengesetz – UKlaG); para. 8 sec. 3 no. 2 Gesetz gegen den unlauteren Wettbewerb (UWG).

13 Para. 15 Gesetz zur Gleichstellung von Menschen mit Behinderungen (Behindertengleichstellungsgesetz – BGG).

14 Baden-Württemberg: Gesetz über Mitwirkungsrechte und das Verbandsklagerecht für anerkannte Tierschutzorganisationen, GBl. 2015, 317; Bremen: Gesetz über das Verbandsklagerecht für Tierschutzvereine, GBl. 2007, 455; Niedersachsen: Gesetz über Mitwirkungs- und Klagerechte von Tierschutzorganisationen, Nds. GVBl. 2017, 108; Hamburg: Gesetz über das Verbandsklagerecht für Tierschutzvereine, HmbGVBl. 2013, 247; Rheinland-Pfalz: Landesgesetz über Mitwirkungsrechte und das Verbandsklagerecht für anerkannte Tierschutzvereine, GVBl. 2014, 44; Schleswig-Holstein: Gesetz zum Tierschutz-Verbandsklagerecht, GVOBl. Schl.-H. 2015, 44; Saarland: Gesetz Nr. 1810 über das Verbandsklagerecht für anerkannte Tierschutzverbände, ABl. 2013, 268.

The merits of the claim are formulated in § 2 (4) UmwRG. According to this, in addition to a breach of legal – not necessarily environmental[15] – requirements that are subject decisions of § 1 UmwRG it is, basically, only required that "the infringement affects matters that are among the objectives the association promotes according to its statutes". Corresponding to the objectives of the Aarhus Convention and its implementing provisions of European law, the representative action under the UmwRG thus ultimately brings judicial control solely on the scale of objective legal requirements – and thus goes well beyond the legal protection by courts in the "normal case". This is undoubtedly in the interest of effective environmental protection. At the same time, however, the fault tolerance with respect to environmental administrative decisions approaches zero and the process risk of the project executing bodies – from the energy industry as well as other affected sectors – is considerable.

1.3 The problem of preclusion

Until a few years ago, the effectiveness (also) of the collective action was subject to a degree of relativization by statutory preclusion rules, which conventionally formed an important instrument in German administrative procedural law. By precluding non-timely submissions in an administrative procedure preceding the judicial process, its exclusion (formally) and, if necessary, a loss of (substantive) law, de facto enforces a procedural and early disclosure on the parties of possible conflicting rights and circumstances. However, by the judgment of 15 October 2015, the ECJ declared § 73 (4) 3 of the Administrative Procedure Act (Verwaltungsverfahrensgesetz – VwVfG) and the referring version of § 2 (3) of the UmwRG,[16] as incompatible with the EIA Directive,[17] and the Industrial Emissions Directive.[18] The European Court of Justice stated

---

15 Sabine Schlacke, op. cit. 9, para. 2 UmwRG rec. 54.
16 The rule was: "If the association has had the opportunity to make observations during the procedure referred to in the first sentence of Paragraph 1 (1), it is prohibited from raising, during the appeal procedure, any objection which it did not raise, or did not raise in good time according to the provisions in force, but which it could have raised during the procedure referred to in the first sentence of Paragraph 1 (1)."
17 Directive 2011/92/EU of the European Parliament and of the Council of 13 December 2011 on the assessment of the effects of certain public and private projects on the environment, OJ 2012 L 26, 13 December 2011, 1, amended by Directive 2014/52/EU, OJ 2014 L 124, 16 April 2014, 1.
18 See already Annette Guckelberger and Frederic Geber, "Präklusion unter dem Einfluss der Aarhus-Konvention", *Zeitschrift für Europäisches Umwelt- und Planungsrecht* (2014), Vol. 12, 171–177; regarding the Scope of the decision and the

that, although "neither Article 11(4) of Directive 2011/92 nor Article 25(4) of Directive 2010/75 excludes an action before an administrative authority preceding the legal proceedings and does not prevent national law from requiring the applicant to exhaust all administrative review procedures before being authorised to bring legal proceedings, those provisions of EU law do not, however, allow restrictions on the pleas in law which may be raised in support of legal proceedings".[19] Rather, those provisions pursue "ensuring broad access to justice in the area of environmental protection. Paragraph 2 (3) of the UmwRG and Paragraph 73 (4) of the VwVfG lay down specific conditions restricting the review by the courts which are not provided for in either Article 11 of Directive 2011/92 or Article 25 of Directive 2010/75. Such a restriction laid on the applicant as to the nature of the pleas in law which he is permitted to raise before the court reviewing the legality of the administrative decision which concerns him cannot be justified by considerations of compliance with the principle of legal certainty. It is in no way established that a full review by the courts of the merits of that decision would undermine that principle".[20]

With regard to the efficiency of administrative procedures in favour of preclusion, the European Court of Justice pointed out that the purpose of the guidelines "is not only to ensure that the litigant has the broadest possible access to review by the courts but also to ensure that that review covers both the substantive and procedural legality of the contested decision in its entirety".[21] As a result of the decision of the European Court of Justice, the federal legislator was forced to change German law. The law passed in 2017 to adapt the Environmental Remedies Act and other provisions to European and international law requirements,[22] included, among others, a deletion of the § 2 (3) UmwRG (old version), which was contested by the European Court of Justice § 7 (4) UmwRG determines since then: "In appeal proceedings against a decision pursuant to § 1 sec. 1 sentence 1 numbers 1 to 2b, § 73 sec. 4 sentences 3 to 6 of the Administrative Procedure Act, also in the cases of its sec. 8, does not apply". For the Environmental Impact Assessment,

---

immediate consequences Attila Széchényi, "EuGH locuta – Präklusion finita? Das Urteil des EuGH vom 15. Oktober 2015 (Rs. C-137/14) und seine Auswirkungen auf Behördenpraxis und Gerichtsverfahren", *Bayerische Verwaltungsblätter* (2016), 366, 368 et seq.

19 European Court of Justice, judgment of 15 October 2015, case C-137/14, ECLI:EU:C:2015:683, *Commission v. Germany*, rec. 76.
20 Ibid., rec. 77 et seq.
21 Ibid., rec. 80.
22 BGBl. I 2017, 1298.

the exclusion of objections contained in § 21 (4) 1 Law on Environmental Impact Assessment (Gesetz über die Umweltverträglichkeitsprüfung – UVPG) was designed as a merely formal preclusion, that is to say only regarding the administrative procedure.[23] However, § 74 (4) 3 VwVfG remained untouched, as the legislator considered "that the exclusion of objections could be retained during the administrative procedure even after the decision of the European Court of Justice".[24] In the case-law of the Federal Administrative Court, the limitation of the preclusion has also found clear expression. The court emphasized the fundamental compatibility of § 74 (4) 3 VwVfG with EU law.[25] Nonetheless, it has unreservedly established that the provision is inapplicable in the light of European Court of Justice's case-law regarding projects in which an environmental impact assessment has to be carried out.[26] Furthermore, the court has ruled that the inapplicability due to European law also relates to the parallel provisions in the administrative procedural laws of the federal states,[27] and that ordering non-compliance in § 7 (4) UmwRG also extends to this.[28] In addition, § 7 (4) UmwRG already precludes the applicability of the preclusive provision of § 73 (4) 3 VwVfG in cases where an EIA requirement can exist if a preliminary examination of the individual case shows the possibility of significant adverse environmental effects.[29] Lastly, "Article 11 (1) of the EIA Directive should not be construed as

---

23 *UVPG – Gesetz über die Umweltverträglichkeitsprüfung Umwelt-Rechtsbehelfsgesetz: Kommentar*, eds. Werner Hoppe, Martin Beckmann and Martin Kment (Köln: Carl Heymanns Verlag, 2018), para. 21 rec. 28 et seq.
24 BT-Drucks. 18/9526, p. 2; see also Jonas Schüren and Malte Kramer, "EuGH-Entscheidung zum UmwRG: Das Aus für materielle Präklusion und traditionelle Verfahrensfolgenlehre?", *Zeitschrift für Umweltrecht* (2016), 400–407.
25 German Federal Administrative Court, judgment of 25 May 2016, case 3 C 2.15, BVerwGE 155, 218 rec. 25.
26 German Federal Administrative Court, judgment of 22 November 2016, case 9 A 25/15, *Neue Zeitschrift für Verwaltungsrecht* (2017), 627 rec. 11; Burghard Hildebrandt and Eva Koch, "Unionsrechtswidrigkeit der Präklusion – Neues aus Leipzig?", *Neue Zeitschrift für Verwaltungsrecht* (2017), 1099–1101.
27 German Federal Administrative Court, judgment of 22 October 2015, case 7 C 15/13, *Neue Zeitschrift für Verwaltungsrecht* (2016), 308 rec. 26; judgment of 30 March 2017, German Federal Administrative Court, case 7 C 17/15, *Neue Zeitschrift für Verwaltungsrecht – Rechtsprechungs-Report* (2017), 685 rec. 20 et seq.
28 German Federal Administrative Court, judgment of 14 December 2017, case 4 C 6/16, *Deutsches Verwaltungsblatt* (2018), 656 rec. 12.
29 German Federal Administrative Court, decision of 29 June 2017, case 9 A 8/16, *Neue Zeitschrift für Verwaltungsrecht* (2017), 1717.

restricting it to the exclusion of the objection of direct environmental considerations. There is no room for a differentiated approach in this respect ... The - possible - environmental effects are, therefore, cause for the rights guaranteed ... by Art. 11 (1) EIA Directive; however, the provision does not contain a restriction relating thereto".[30] Even though the (material) preclusion was not completely eliminated, it is only of minor importance in the context of the association's claim under the UmwRG. It still applies to non-EIA-compliant projects. However, those projects in which the effects of occlusion were of particular importance due to their complexity tend not to be covered.[31] Planning assessments of projects subject to EIA, which include in particular the grid expansion in the field of energy, are therefore subject to a considerably higher process and cancellation risk.[32] As a result, public participation, which is often planned in the context of network expansion, is potentially losing significance.[33]

2 Proposal for a directive on representative actions for the protection of the collective interests of consumers

The possibilities of representative actions (also) in the energy sector will be extended considerably in the future. In April 2018, the EU Commission submitted a proposal on representative actions for the protection of the collective interests of consumers and repealing Directive 2009/22/EC.[34] At a first reading, the European Parliament

---

30 German Federal Administrative Court, judgment of 30 March 2017, case 7 C 17/15, *Neue Zeitschrift für Verwaltungsrecht - Rechtsprechungs-Report* (2017), 685 rec. 22, 26.
31 Cf. German Federal Constitutional Court, decision of 18 September 2017, case 1 BvR 361/12, *Neue Zeitschrift für Verwaltungsrecht* (2018), 406 rec. 14.
32 Alexander Stefan Rieger and Johannes Groß, "Wegfall der Präklusion in UVP-Verfahren. Gesteigertes Investitionsrisiko durch EuGH-Urteil", *Neue Zeitschrift für Baurecht und Vergaberecht* (2017), 199; cf. Thomas Jacob, "Subjektiver Rechtsschutz, Einwendungspräklusion und Bestandskraft - klare Worte vom EuGH? ", *Juris Monatszeitschrift* (2016), 166, 167 et seq.
33 Burghard Hildebrandt and Eva Koch, op. cit., 1101; Thorsten Siegel, "Die Präklusion im europäischen Verwaltungsrecht", *Neue Zeitschrift für Verwaltungsrecht* (2016), 341; Christian Zeissler and Vera Schmitz, "Das Urteil des EuGH vom 15.10.2015 in der Rechtssache C-137/14 als Ende der umweltrechtlichen Präklusion?", *Umwelt- und Planungsrecht* (2016), 3; different views on public participation, see Sabine Weidermann, "Wegfall der Präklusion - Zum praktischen Wert der Öffentlichkeitsbeteiligung", *Die Öffentliche Verwaltung* (2017), 939–943.
34 European Commission, Proposal for a Directive of the European Parliament and of the Council on representative actions for the protection of the collective

voted in favour of more than 100 amendments, but in principle approved the proposal.³⁵ At the end of November 2019, the Council agreed on a general approach and proposed significant modifications.³⁶ A political agreement could be reached in the trilogue between the co-legislators in June 2020.³⁷ Based on this and the subsequent political agreement on the Council's position at first reading in September 2020, the General Secretariat of the Council has submitted a draft for this position in October 2020.³⁸ Regarding the fact that the European Parliament's Committee on Legal Affairs will recommend the European Parliament, at second reading, to approve the Council's position at first reading without amendment,³⁹ it is very

---

interests of consumers, and repealing Directive 2009/22/EC, COM(2018) 184 final of 11 April 2018; for more information Susanne Augenhofer, "Die Reform des Verbraucherrechts durch den 'New Deal' – ein Schritt zu einer effektiven Rechtsdurchsetzung?", *Europäische Zeitschrift für Wirtschaftsrecht* (2019), 5–13; Tanja Domej, "Die geplante EU-Verbandsklagerichtlinie – Sisyphos vor dem Gipfelsieg?", *Zeitschrift für Europäisches Privatrecht* (2019), 446–471; Axel Halfmeier and Peter Rott, "Verbandsklage mit Zähnen? – Zum Vorschlag einer Richtlinie über Verbandsklagen zum Schutz der Kollektivinteressen der Verbraucher", *Verbraucher und Recht* (2018), 243–250; Tobias B. Lühmann, "Der Vorschlag einer europäischen Verbandsklage", *Neue Juristische Wochenschrift* (2019), 570–575; concerning the background, see also Caroline Meller-Hannich and Elisabeth Krausbeck, "Modelle und Praxis des kollektiven Rechtsschutzes in europäischen Mitgliedstaaten", *Deutsches Autorecht Extra* (2018), 725–729.

35 European Parliament Legislative Resolution TA/2019/0222 of 26 March 2019 on the proposal for a directive of the European Parliament and of the Council on representative actions for the protection of the collective interests of consumers, and repealing Directive 2009/22/EC.

36 Council of the European Union, Proposal for a Directive on representative actions for the protection of the collective interests of consumers, and repealing Directive 2009/22/EC, 14600/19 of 28 November 2019.

37 Directive of the European Parliament and of the Council on representative actions for the protection of the collective interests of consumers, and repealing Directive 2009/22/EC – Political agreement, 9592/20 of 16 July 2020.

38 Position of the Council at first reading with a view to the adoption of a Directive of the European Parliament and of the Council on representative actions for the protection of the collective interests of consumers and repealing Directive 2009/22/EC, 9573/20 of 21 October 2020.

39 Draft Directive of the European Parliament and of the Council on representative actions for the protection of the collective interests of consumers and repealing Directive 2009/22/EC (first reading) – Decision to use the written procedure for the adoption of the Council's position at first reading and of the statement of the Council's reasons, 11846/20 of 23 October 2020, fn. 5.

likely that the draft represents the final text. In the following, some essential aspects of the discussed innovations are presented.

2.1 Scope

The scope of the directive is to be extended significantly in comparison with the previous regulation, which still applies to consumer protection law in the narrower sense.[40] The explanatory memorandum of the Commission states: "The scope of the Directive will be expanded to cover other horizontal and sector-specific EU instruments relevant for the protection of collective interests of consumers in different economic sectors such as financial services, energy, telecommunications, health and the environment. This amendment would make the procedure more responsive to the broad spectrum of infringements in economic sectors where the traders' illegal practices may affect a large number of consumers".[41] Specifically, the scope of application of the new rules is defined in Article 2 (1) in conjunction with Annex I to the Draft Directive. Consumer protection-oriented representative actions in the energy sector should, according to the Commission proposal, be admissible in the future with regard to possible infringements of specific legal acts. These acts include the Directive 2009/72/EC of the European Parliament and of the Council of 13 July 2009 concerning common rules for the internal market in electricity and repealing Directive 2003/54/EC, the Directive 2009/73/EC of the European Parliament and of the Council of 13 July 2009 concerning common rules for the internal market in natural gas and repealing Directive 2003/55/EC, and the Directive 2009/125/EC of the European Parliament and of the Council of 21 October 2009 establishing a framework for the setting of eco-design requirements for energy-related products. They also include the Regulation (EC) No 1222/2009 of the European Parliament and of the Council of 25 November 2009 on the labelling of tyres with respect to fuel efficiency and other essential parameters, the Directive 2010/31/EU of the European Parliament and the Council of May 19th, 2010 on the energy performance of buildings, the Directive 2012/27/EU of the European Parliament and of the Council of 25 October 2012 on energy efficiency,

---

40 Cf. Article 1 in conjunction with Annex I Directive 2009/22 of the European Parliament and of the Council of 23 April 2009 on injunctions for the protection of consumers' interests, OJ L 110, 1 May 2009, 30.
41 Proposal for a Directive of the European Parliament and of the Council on representative actions for the protection of the collective interests of consumers, and repealing Directive 2009/22/EC, COM(2018) 184 final, pp. 2 et seq.

amending Directives 2009/125/EC and 2010/30/EU and repealing Directives 2004/8/EC and 2006/32/EC, and finally the Regulation (EU) 2017/1369 of the European Parliament and of the Council of 4 July 2017 setting a framework for energy labelling and repealing Directive 2010/30/EU.

The European Parliament has also spoken in favour of including Regulation (EC) No 715/2009 of the European Parliament and of the Council of 13 July 2009 on conditions for access to the natural gas transmission networks and repealing Regulation (EC) No 1775/2005.[42] The Council rejected this as well as the inclusion of Directive 2010/31/EU. In addition, it proposed to include the national transposition law and to clarify the covered rules. These are, Directive 2009/72/EC: Article 3 and Annex I, Directive 2009/73/EC: Article 3 and Annex I, Directive 2009/125/EC: Article 14 and Annex I, Regulation (EC) No 1222/2009: Articles 4–6, Directive 2012/27/EU: Articles 9–11a, and Regulation (EU) 2017/1369: Articles 3–6. Thus, the Directive will, in the future, capture material secondary legislation in the areas of the internal energy market and energy efficiency. The representative action should not be comprehensive in this regard, but according to the Commission's proposal in relation to "infringements by traders ... that harm or may harm the collective interests of consumers"[43]. This benchmark is not very concrete and hardly suitable for the area of energy law to provide a clear definition of the area open to representative action. The amendment proposed by the European Parliament is scarcely any more clear. After that, the directive should "apply to representative actions brought against infringements with a broad consumer impact by traders of provisions of the Union law listed in Annex I that protect the collective interests of consumers"[44]. In view of the legally enforceable applicability of the rules, therefore, it is absolutely necessary to add the concretisations proposed by the Council. The Commission and the European Parliament have accepted those clarifications in the trilogue negotiations.

2.2 Design

Article 5 of the draft Directive sets out in detail the conditions for collective actions covered by the scheme. The text of the Commission proposal reads as follows: "1. Member States shall ensure that representative actions can be brought before national courts or administrative authorities by qualified entities provided

---

42 Amendment 108.
43 Proposal for a Directive COM(2018) 184 final, op. cit., p. 13.
44 Amendment 33.

that there is a direct relationship between the main objectives of the entity and the rights granted under Union law that are claimed to have been violated in respect of which the action is brought. 2. Member States shall ensure that qualified entities are entitled to bring representative actions seeking the following measures: a) an injunction order as an interim measure for stopping the practice or, if the practice has not yet been carried out but is imminent, prohibiting the practice; b) an injunction order establishing that the practice constitutes an infringement of law, and if necessary, stopping the practice or, if the practice has not yet been carried out but is imminent, prohibiting the practice. In order to seek injunction orders, qualified entities shall not have to obtain the mandate of the individual consumers concerned or provide proof of actual loss or damage on the part of the consumers concerned or of intention or negligence on the part of the trader. 3. Member States shall ensure that qualified entities are entitled to bring representative actions seeking measures eliminating the continuing effects of the infringement. These measures shall be sought on the basis of any final decision establishing that a practice constitutes an infringement of Union law listed in Annex I, harming collective interests of consumers, including a final injunction order referred to in paragraph (2)(b). 4. Without prejudice to Article 4(4), Member States shall ensure that qualified entities are able to seek the measures eliminating the continuing effects of the infringement together with measures referred to in paragraph 2 within a single representative action".

In addition to a deletion of sec. 3 sentence 2 and sec. 4,[45] as well as editorial amendments, the European Parliament has proposed the replacement of the term "qualified entities" by "qualified representative entities". Furthermore, paragraph 1 shall be supplemented by the following subparagraphs: "The qualified representative entities are free to choose any procedure available under national or Union law ensuring the higher level of protection of the collective consumer interest. Member States shall ensure that no other ongoing action has been brought before a court or an administrative authority of a Member State regarding the same practice, the same trader and the same consumers." The Council has proposed the following: "1. Member States shall ensure that representative actions in accordance with Article 2 can be brought before national courts or administrative authorities by qualified entities. When bringing a representative action, the qualified entity shall provide to the court or to the administrative authority sufficient information on the consumers concerned

---

45 Amendment 57–58.

by the action. 2. Member States shall ensure that qualified entities are entitled to seek, as appropriate, at least the following measures: (a) injunction measures, (b) redress measures. 4. Member States may enable qualified entities to seek, as appropriate the measures referred to in paragraph 2 within a single representative action. Member States may provide that these measures are issued within a single decision. 4a. Member States shall ensure that, within a representative action, consumers' interests are represented by qualified entities and that the entities have the rights and obligations of a party to the proceedings. The consumers concerned by the action shall be entitled to benefit from the measures referred to in paragraph 2. The consumers concerned by the action may bear costs of the proceedings only in exceptional circumstances in accordance with national law. Member States may set out rules allowing for modest entry fees or similar participation charges." Injection and redress matters should also be designed in detail in the new Articles 5a and 5b according to the Council.

The requirements for qualified (representative) entities, which in the future will also be eligible in the energy sector, are set out in Article 4 of the draft Directive.[46] The text of the Commission proposal reads as follows: "1. Member States shall ensure that representative actions can be brought by qualified entities designated, at their request, by the Member States in advance for this purpose and placed in a publicly available list. Member States shall designate an entity as qualified entity if it complies with the following criteria: a) it is properly constituted according to the law of a Member State; b) it has a legitimate interest in ensuring that provisions of Union law covered by this Directive are complied with; c) it has a non-profit making character. 2. Member States shall assess on a regular basis whether a qualified entity continues to comply with these criteria. Member States shall ensure that the qualified entity loses its status under this Directive if it no longer complies with one or more of the criteria listed in the first subparagraph. . . . 3. Member States shall ensure that in particular consumer organisations and independent public bodies are eligible for the status of qualified entity. Member States may designate as qualified entities consumer organisations that represent members from more than one Member State. . . .". The European Parliament considers that, in addition to clarifications, it should be further standardized that at least one qualified (representative) entity should be designated in each Member State and that a list of such bodies should be maintained and published at Member State level. The latter should also be done

---

46 Cf. critical Jürgen Basedow, "Trippelschritte zum kollektiven Rechtsschutz", *Europäische Zeitschrift für Wirtschaftsrecht* (2018), 612.

by the Commission. In addition, further requirements are to be made for the facilities: "(c a) it acts in a way that is independent of other entities and from persons other than consumers who might have an economic interest in the outcome of the representative actions, in particular from market operators; (c b) it does not have financial agreements with plaintiff law firms beyond a normal service contract; (c c) it has established internal procedures to prevent a conflict of interest between itself and its funders".

The Council has proposed further differentiation. Article 4 contains requirements for qualified entities for the purpose of domestic representative actions: "1. Member States shall ensure that domestic representative actions can be brought by qualified entities designated, at their request, by the Member States for this purpose. 3. Member States shall ensure that in particular consumer organisations, including those representing consumers from more than one Member State, are eligible to apply for the status of qualified entity in accordance with national law. 4a. Member States may designate public bodies as qualified entities. 4b. Member States may designate a qualified entity, at its own request, on an ad hoc basis for the purpose of a particular representative action. 4c. Information on qualified entities designated in advance for the purpose of any representative action shall be made available to the public." Partly divergent, extensive requirements should, in the Council's view, be laid down in Article 4a with regard to the designation of qualified entities for the purpose of cross-border representation actions. In addition, a provision is to be created with Article 4b, which explicitly addresses the bringing of cross-border representative actions.

In the trilogue negotiations a new version of the provision has been agreed upon, mixing the concepts discussed before. In its probably final wording Article 4 reads: "1. Member States shall ensure that representative actions can be brought by qualified entities designated by the Member States for this purpose. 2. Member States shall ensure that entities, in particular consumer organisations, including those representing members from more than one Member State, are eligible to be designated for the status of qualified entity for the purpose of bringing domestic representative actions or cross- border representative actions, or both. 3. Member States shall designate an entity, at its request, as a qualified entity for the purpose of bringing cross-border representative actions, if it complies with all of the following criteria: (a) it is a legal person properly constituted according to the law of the Member State of its designation and can demonstrate 12 months of actual public activity in the protection of consumer interests prior to its designation request; (b) its statutory purpose demonstrates that it has a legitimate interest in protecting consumer interests as provided by Union law covered by this Directive; (c) it has a non-profit making character; (d) it is not subject to insolvency procedure or declared insolvent; (e) it is independent and not influenced by persons, other than consumers, who have an economic interest in the bringing of any representative action, in particular by traders, including in case of funding by third parties, and, to that end, has established procedures preventing such influence as well as conflict of interest between itself, its funders and consumer interest; (f) it discloses publicly by any appropriate means, in particular on its website, in plain and intelligible language, information demonstrating compliance with the above listed criteria and information about the sources of its funding in general, its organisational, management and membership structure, objectives and activities. 4. Member States shall ensure that the criteria they use to designate an entity as a qualified entity for the purpose of bringing domestic representative actions are consistent with the objectives of this Directive to make effective and efficient functioning of such actions. 5. Member 1.

States may decide that the criteria set out in paragraph 3 apply also to the designation of qualified entities for the purpose of bringing domestic representative actions. 6. Member States may designate a qualified entity, at its own request, on an ad hoc basis for the purpose of bringing a particular domestic representative action, if it complies with the criteria for qualified entities as provided for in national law. 7. Notwithstanding paragraphs 3 and 4, Member States may designate public bodies as qualified entities for the purpose of bringing representative actions. Member States may provide that public bodies already designated as qualified entities in the meaning of Directive 2009/22/EC shall remain eligible as qualified entities." This provision will be accompanied by two articles dealing with information on and monitoring of qualified entities and cross-border representative actions.

2.3 Consequences for the energy sector

Although it remains to be seen if the legislation will be finalized at the end of the legislative procedure exactly according to the draft of the Council's position representing the results of the trilogue negotiations, it is already foreseeable that effective national laws for representative actions will need to be created. At the same time, the requirements for suit-entitled associations are designed in such a way that their existence is ensured.[47] Ultimately, a clear orientation of

---

47 Axel Halfmeier and Peter Rott, cit., 250, criticise an insufficiently clear design, at least for a transitional period.

the consumer protection-related legal protection by associations based on the model of environmental law can be stated. For the energy sector, future regulation means that representative actions will be extended to almost all areas with relevance for consumers.

Conclusions

Already on the basis of the applicable law, the representative action is of great importance in the energy sector. Although its scope has been limited so far to environmental law in both German and EU law, numerous infrastructure projects in the energy industry can be attacked and, if necessary, prevented in this way. If the directive proposed by the Commission on representative actions for the protection of the collective interests of consumers is adopted by the European legislator – as the previous opinions of the European Parliament and of the Council suggest, irrespective of the proposed amendments – the scope of the representative action in the energy sector will be considerably extended in the future. The restriction on consumer protection interests laid down in the proposed directive leads to the fact that not every potential violation of the energy law requirements can be objected to by associations. However, whether the new rules will provide sufficient indications as to how to differentiate between consumer and non-consumer protection requirements of EU energy law depends on whether the Council will prevail with its proposals in this regard. Regardless of the final design, however, it is to be expected that the extension of the representative action in the energy sector will further increase the process risks. At the same time, the pressure is increasing on the energy industry to comply with EU environmental and consumer energy law.

Dr. Athina Moraiti*

15 Consumer protection and the right to out-of-court dispute settlement in the oil & gas sector. Regulator or Ombudsman model for Energy Alternative Dispute Resolution in Europe?

Abstract: The Third Energy Package and the Energy Efficiency Directive as well as horizontal consumer rights legislation lay out a number of consumer rights, which are already applicable in the energy sector. However, these laws have not yet had the necessary impact, which is to empower and protect European consumers of oil and gas. This chapter analyses the existing rights for consumers in oil and gas markets, identifies the gaps and tries to find out which entity would be the most appropriate to apply the Alternative Dispute Resolution for Europe. It is common that the regulatory authorities most of the times include in their structure special bodies in order to resolve any dispute arisen. This issue might cause doubts about the independence of the regulator for the implementation of the directive, as well as for the actual protection of the consumer. The recent European Parliament legislative resolution on the establishment of common rules for the internal market in electricity requires the Member States to ensure that final customers have access to an independent mechanism, in order to settle disputes concerning the rights and obligations established under the Directive. Which model is the best for the European Union to follow? Should we establish an ADR section in a European Commission Level?

Keywords: Consumer protection, oil and gas energy disputes, ombudsman, mediator-arbitration, regulators as ADR bodies

1 Consumer rights and protection

Even though the Third Energy Package and the Energy Efficiency Directive provide for consumer rights which are already applicable in the electricity sector,

---

* Dr. Athina Moraiti, Greek Ministry of Finance, General Secretariat of Public Property. The opinions expressed in this publication are those of the author. They do not purport to reflect the opinions or views of the Ministry of Finance or its employees.

this is not the case for the energy rights for oil and gas consumers.[1] This chapter attempts to analyse the existing rights for the consumers in oil and gas markets, picks out the differences and gaps between electricity and oil and gas rights, and emphasizes on the future challenges, having as a goal to investigate which entity meets the best criteria to apply the Alternative Dispute Resolution (hereinafter ADR) for Europe. It is not questionable that energy services are more than necessary for people. Nowadays we use energy in our everyday lives for almost all of our activities. Some would say that while electricity is characterized by EU Directive 2009/72/EC as a universal service, it is not the case for gas, as the relevant Directive does not contain the word 'universal'.[2] Does this mean that gas is not of the same importance as electricity? According to Eurostat's research, gas consumption has increased by almost 3,7 %. Gas is, therefore, also an essential energy source for European households.[3] Article 3 of the Directive 2009/73,[4] provides for the obligation of Member States to take the appropriate measures to protect final customers. The Member States have to ensure that

---

1   Consumer Rights Directive (Directive 2011/83/EU of the European Parliament and of the Council of 25 October 2011 on consumer rights, amending Council Directive 93/13/EEC and Directive 1999/44/EC of the European Parliament and of the Council and repealing Council Directive 85/577/EEC and Directive 97/7/EC of the European Parliament and of the Council), Unfair Contract Terms Directive (Directive 93/13/EEC of the Council of 5 April 1993 on unfair terms in consumer contracts) and Unfair Commercial Practices Directive (Directive 2005/29/EC of the European Parliament and of the Council of 11 May 2005 concerning unfair business-to-consumer commercial practices in the internal market and amending Council Directive 84/450/EEC, Directives 97/7/EC, 98/27/EC and 2002/65/EC of the European Parliament and of the Council and Regulation (EC) No 2006/2004 of the European Parliament and of the Council).
2   Directive 2009/72 of the European Parliament and of the Council of 13 July 2009 concerning common rules for the internal market in electricity and repealing Directive 2003/54/EC lays down electricity as a universal service meaning that all household consumers should have "the right to be supplied with electricity of a specified quality within their territory at reasonable, easily and clearly comparable, transparent and non-discriminatory prices".
3   According to Eurostat in 2017, gross inland consumption of natural gas in the EU-28 increased by 3.7 % compared with 2016. In 2012 natural gas consumption covered 39 % of the total energy needs of households in 2010. The share of electrical energy was 24 % in 2010 compared to 21 % in 2000.
4   Directive 2009/73/EC of the European Parliament and of the Council of 13 July 2009 concerning common rules for the internal market in natural gas and repealing Directive 2003/55/EC.

the suppliers are in a position to avoid disconnections and that the system of transport is functional to the maximum possible extent. It is an obligation for every member state in order to promote energy efficiency, to set the legal framework and define the concept of vulnerable customers and inform them about the risk of potential disconnection and provide alternative solutions and, thus, set an access price which can be affordable for the consumers. The ultimate scope of the internal energy market is to give every consumer the right to be able to choose their preferred supplier irrespective of their place of residence. Even though the European energy market has made steps towards the completion and functionality of the internal energy market, after three legislative packages, European consumers have not yet enjoyed the benefits of truly competitive energy markets.[5] The Third Energy Package has provided for the right of the consumers to choose their supplier irrespectively of his origin.

Since consumers are at the core of the EU' s energy policy, the Directives set a wide range of initiatives to constitute them a more active part of the clean energy transition by providing them with the necessary information and offering them options on the ways they can participate in the energy market. Having access to valuable information about their consumption, or about the cost of their connection to the network, consumers will be better protected, will acquire energy conscience and thus will get a stronger position in the energy supply chain. The EU legislation guarantees that all EU residents enjoy general consumer rights, but those rights have to be clearly set out in the member states' national law. In order to help consumers understand the existing offers in the electricity and gas sectors, the European Commission established the Citizens' Energy Forum in 2008 that meets annually to explore the perspectives for the consumers and their protection in the EU energy market.[6] Transparency on applicable prices and tariffs is very important for consumers. For this reason, the Third Energy Package gives emphasis on this matter as well as on standard terms and conditions of the energy contracts.[7] As regards the information requirements that need to be met in energy contracts, these are provided in Articles 5 (for on-premises contracts) and 6 (for distance and off-premises contracts) of the Directive 2011/83/EU (Consumer Rights Directive). These

---

5 "Consumer rights in electricity and gas markets", *BEUC position paper*, December 2013, https://www.beuc.eu/publications/x2013_083_mst_consumer_rights_in_electricity_and_gas_markets.pdf.
6 European Commission, "The 11th Citizens Energy Forum", September 2019, https://ec.europa.eu/info/events/11th-citizens-energy-forum-2019-sep-12_en.
7 See Annex I of Directives 2009/72 and 2009/73 about the rights of gas consumers.

Articles apply when needed, together with what Annex I of the Directives 2009/72 and 2009/73 provides. In its recent decision, the Court held that the aim of the Directive 2009/73 is to pursue the achievement of an internal market in natural gas that is entirely and effectually open and competitive and in which all consumers can freely choose their suppliers, and all suppliers can freely supply their products to their customers, which means that the price of the supply of natural gas must be fixed exclusively by the interplay of supply and demand.[8] In the case of *Repsol Butano SA (C-473/17), DISA Gas SAU (C-546/17) v. Administración del Estado*,[9] the ECJ held that the condition of proportionality laid down in Article 15(3)(c) of Directive 2006/123 must be interpreted as not precluding measures which set a maximum price for bottled LPG and which require certain operators to carry out home delivery of that gas, provided that those measures are maintained only for a limited duration and do not go beyond what is necessary in order to achieve the objective of general economic interest pursued. European consumer protection law only applies to B2C (business to consumer) relationships, not B2B (business to business) or P2P (peer to peer) relationships – central to the collaborative economy. If the supplier is qualified as a 'trader' the rules will apply; if, on the contrary, the supplier is qualified as a 'prosumer',[10] their relations with consumers will be governed only by the traditional contract and/or tort law.

---

8   European Court of Justice, judgment of 7 September 2016, case C-121/15, EU:C:2016:637, *ANODE v. Premier ministre and Others*, para. 26.
9   European Court of Justice, judgment of 11 April 2019, joined cases C-473/17 and C-546/17, ECLI:EU:C:2019:308, *Repsol Butano SA and DISA Gas SAU v. Administración del Estado*.
10 See European Commission, Best practices on Renewable Energy Self-consumption, Commission Staff Working Document accompanying the document Communication from the Commission to the European Parliament, the Council, the European Economic and Social Committee and the Committee of the Regions Delivering a New Deal for Energy Consumers, SWD(2015) 141 final of 15 July 2015, 1. Introduction: "Thanks to technology development and innovation driven by EU and national policies, over the last few years we have seen the realization of effective renewable energy technologies, for both large and small-scale use, alongside considerable cost reductions. As a result, businesses and households can increasingly produce and consume, some or all, their own electricity, either instantaneously or in a deferred manner through decentralized storage, behind the connection point with the grid (i.e. the meter). Through the process of 'self-consumption', passive consumers are therefore becoming active 'prosumers' (i.e. producers and consumers of renewable energy)."

This creates a potentially very important gap in the field of consumer protection in energy contracts, as, in the cases where the providers of the services are not themselves traders (they are small producers for example) the consumers will not be able to turn against the prosumer to restore their damage. In such claims, the claimant will not be able to use the favourable rules on jurisdiction and applicable law foreseen in favour of consumers,[11] but according to the civil procedure rules, they will also have to bear the burden of proof and might violate a contractual term, or fault. And what is worse for the protection of the consumers is that the defendant in such claims, may be unable to offer any refund, to the consumer. Inspired by the practices of several Member States, the Commission in its 2016 Communication suggested the use of thresholds as criteria for the qualification of a 'trader': a) the frequency of services, i.e. whether the services are offered regularly or on a purely marginal or accessory basis, b) the motive for profit, as opposed to the objective of exchanging assets or skills, c) the income generated by the activity concerned, and if it is higher/lower than those of other activities undertaken by the same person. It is important to note that these criteria only serve to clarify cases at the two extremes of the spectrum of possibilities; they remain of little use in most routine situations and would deserve further clarification.[12] These criteria have to be applied for the protection of the energy prosumers, who do not, of course, exist in the oil & gas sector. Recently the European Parliament approved the new rules on consumers in energy markets and the broader electricity market reform and clean energy package. Among other rights, consumers will have the possibility to actively participate in the electricity markets and become prosumers. They will be able to switch supplier within three weeks, and they will be able to compare the prices and receive information about energy consumption and costs. Naturally, these rights will have to be applied in the oil & gas sector. The recent directive (EU) 2019/944 of June 5, 2019, on common rules for the internal market for

---

11 Under the Rome I Regulation [Regulation (EC) 593/2008 of the European Parliament and of the Council of 17 June 2008 on the law applicable to contractual obligations (Rome I), OJ L 177, 4 July 2008, 6] and the Brussels I Regulation [Regulation (EU) 1215/2012 of the European Parliament and of the Council of 12 December 2012 on jurisdiction and the recognition and enforcement of judgments in civil and commercial matters (Brussels I Regulation (recast)), OJ L 351, 20 December 2012, 1].

12 See Vassilis Hatzopoulos and Sofia Roma, "Caring for Sharing? The Collaborative Economy under EU law", *Common Market Law Review* (2017), 81–128; and more extensively see Vassilis Hatzopoulos, "The Collaborative Economy and EU Law" (Oxford: Hart Publishing, 2018).

electricity and amending Directive 2012/27/EU empowers the consumers and their protection. The increasing globalization of the economy at the global level and the smooth functioning of the single market at EU level had made, the previous years, imperative the need for unification or harmonization, at least, of those areas of private law, which were crucial for a stumble movement of goods, individuals, services and capital. Legislative and administrative harmonization in certain areas of civil law and public law was in many ways useful and desirable, perhaps necessary in order to fully cover the area of the consumer's protection at a national level.

In Greece, the characterization of energy activities as utilities, puts them under the supervision of the Greek State and this incorporates them into the realm of legal interest of public (administrative) law so that the disputes arising between the public authorities and market participants are examined by the administrative courts. Consumer protection in energy is established explicitly in law 4001/2011. The fact that the Greek legislator had rational and crystal disposal to protect energy consumers is very important, as their protection is also a matter of public interest. The law provides for clarification and definition of the term 'energy poverty' which is "the state of consumers who are in a difficult economic state", and also gives a clear definition of Vulnerable Customers who are classified in five different groups.[13] The Energy Regulatory Authority (RAE) in Greece controls and monitors the compliance with the obligations provided for Vulnerable Customers and has the responsibility to impose sanctions under the provisions of the Article 36 about administrative penalties if there is a violation of these obligations. By decision of the Energy Regulatory Authority, issued after the hearing of the two parties, a fine can be imposed on Owners and Managers of Electricity grids and Gas Companies engaged in Energy Activities, in case of violations of the provisions of law 4001/2011. In such a case, fines up to 10 % of their annual turnover can be imposed. The fine is proportional to the severity and frequency of the violation. The Greek Energy Regulatory Authority (RAE) also supervises the implementation of measures to protect the rights of the Customers. Article 3 of the law 4001/2011 provides, among others, that the Energy Regulatory Authority has the competence to protect the interests of customers, and particularly of the Vulnerable Customers, especially with regard to prices, transparency of tariffs and charges, terms of energy supply, the security of supply, regular supply, quality of service provided and the effective exercise of customers' right to choose a supplier. RAE, by decision,

---

13 Greek Law 4001/2011, Article 52.

can oblige the Transmission System Operator to provide consumption data to the Customers, without charge.[14] Furthermore, according to Article 24, RAE only deals with Customer Complaints if they arise from, or relate to, regulatory oversight matters provided for in law 4001/2011 and are specified by the regulatory decisions issued by RAE. RAE does not deal with issues relating to civil or commercial disputes.[15]

2 Oil & gas industry disputes

The oil & gas industry is one of the most vibrant and dynamic industries in the world.[16] This dynamism and fast evolution and the fact that it contains a very big range of activities and processes, might lead to disputes.[17] The big range of activities, the number of the people involved, as well as the complexity of the contracts, often end up in difficulty to solve disputes.[18] The difficulty for the resolution of the disputes lies in economic, political and environmental factors.[19] A dispute can be defined as a disagreement concerning a matter of law, fact, or policy where a claim of one party is met with refusal, or counter-claim by another.[20] As the field of oil & gas industry is international, so are the arisen disputes. Most of the contracts concerning the oil & gas industry have the following characteristics: they are long term and they involve multiple stakeholders, so they can sometimes be complex, both from a technical and legal aspect. Oil contractual disputes might have an economic and reputational impact for a company. Sometimes resolving amicably a dispute, can cost millions. For this reason, both parties in the oil & gas sector, when an issue

---

14 Ibid., Article 24.
15 Ibid.
16 Tim Boykett, Marta Peirano and Simone Boria et al., "Oil Contracts, How to Read and Understand a Petroleum Contract" (Linz, Austria: Times Up Press, 2012).
17 Alphonsus J. Fagan, "An introduction to the petroleum industry" (Government of Newfoundland and Labrador, Department of Mines and Energy, November 1991).
18 Timothy Martin, "Dispute Resolution in the International Energy Sector: An Overview", *Journal of World Energy Law and Business* (2011), 332–368, http://jwelb.oxfordjournals.org.
19 Christian O. H. Wolf, "The Petroleum Sector Value Chain", *University of Cambridge – Judge Business School*, 30 June 2009, https://papers.ssrn.com/sol3/papers.cfm?abstract_id=1514593.
20 Mohammad Alramahi, "Dispute Resolution in Oil and Gas Contracts", *International Energy Law Review* (2011), 78–85, https://ssrn.com/abstract=2159702.

arises, and until it is resolved, have no desire to stop their activities, and once the dispute is resolved, they wish to continue their commercial relationships.

3 Types of disputes – under examination only disputes between individuals and oil & gas companies

When do we have a dispute? A dispute is a situation in which the two contracting parties have clearly opposite views concerning the question of the performance or non-performance of the principal agreement between the parties. Many situations may cause a dispute, such as a delay in the delivery of the equipment, or environmental accidents, or other unexpected incidents. Especially in the oil & gas sector, we distinguish four main categories of disputes: – disputes between countries – disputes between companies against countries – disputes between companies and disputes between individuals and companies.[21] The situations where individuals initiate claims against oil & gas companies are specific: a) personal injury – usually, the harmed person or entity begins a tort claim against the company.[22] b) disputes between investors and states, during a contract, c) when agents or consultants of the oil & gas company after contracting with a government, demand their payment.[23] A further type of disputes arising from oil & gas contracts includes among others, disputes among operators, non-operators and joint ventures in property acquisition, exploration developments, supply and marketing arrangements and construction projects.

4 Methods of dispute resolution

When signing a binding international agreement, the parties may use one of many different dispute resolution methods.[24] These methods contain negotiations, mediation, expert determination, dispute review boards, litigation and arbitration. There are Member States of the EU where their energy ADR body operates within the National Regulatory Authority. This issue might cause

---

21 Timothy Martin, cit., 322–368.
22 For more details on these claims worldwide, see Jonathan Drimmer, "Human Rights and the Extractive Industries: Litigation and Compliance Trends", *Journal of World Energy Law & Business* (2010), 121–139, https://doi.org/10.1093/jwelb/jwq001.
23 Timothy Martin, cit., 322–368.
24 "The Inside Track: How Blue-Chips are Using ADR", *Herbert Smith Freehills*, November 2007, http://sites.herbertsmithfreehills.vuturevx.com/20/10753/landing-pages/6398-adr-report-d4.pdf.

doubts about the independence of the regulator for the implementation of the directive, as well as for the actual protection of the consumer. Even though the Member States usually have a regulatory authority to deal with energy complaints, the model adopted in every Member State is not the same. In the EU and EFTA Member States, a wide range for dispute resolution exists: in Austria, Belgium, Bulgaria, France and Germany we have independent energy ADR bodies;[25] while in Croatia, Cyprus, Estonia, Finland, Hungary, Iceland, Italy, Latvia, Malta, Norway, Holland, Portugal, Spain and Romania we have special ADR bodies within the Regulatory Authorities. After almost 25 years of the liberalization of the energy market, and the enforcement of the Third Energy Package, every Member State is at a different stage of its implementation. In Germany, the Energy Conciliation Board handles disputes between consumers and energy providers (electricity and gas) that concern, for example, the payment of an invoice or the switch to a new energy supplier.

Furthermore, energy suppliers are obligated to take part in ADR procedures if the dispute covers the energy or gas supply. Some energy suppliers also offer additional services like water supply. Related disputes are covered by the General Consumer Conciliation Body.[26] In Poland since 2011, the Information Point for Fuel and Energy Customers operates within the structure of the National Regulatory Authority. The information point supports customers mostly by providing them with information on their rights and obligations in their relations with energy undertakings. In 2017, customers filed a total of 3,417 requests; among them, the most common problems referred to the electricity (65 %) and gas (14 %) sector. Requests focused on issues related to terms and conditions of concluded agreements (23 %) and unfair market practices, especially in the context of switching supplier (25.8 %), switching supplier itself (6 %), and complaints related to invoices (6 %) and measurement systems (3 %) and customer service (3 %).[27]

---

25 According to the European Consumer Center of Germany in "ADR bodies in the EU", 11 March 2020, https://www.evz.de/en/shopping-internet/alternative-dispute-resolution/adr-bodies-in-the-eu.html.

26 For more information see the site of the German Energy Conciliation Body Association "Verein Schlichtungsstelle Energie", https://www.schlichtungsstelle-energie.de/ihr-recht.html.

27 See Annual Report of the National Polish Energy Regulatory Authority, http://www.ure.gov.pl/en/about-us/reports/67,Reports.html.

5 Ombudsman – mediation

Mediation comes from the Latin word 'meditari' which means to concentrate and is a form of alternative dispute resolution (ADR), a way of resolving disputes between two or more parties (in the field of energy, between countries, companies and customers as mentioned above). Mediation is frequently used in common law jurisdictions, but lately, it is also spreading in European and civil law jurisdictions, as a fast, effective and affordable tool. A third party, the mediator, assists the parties to negotiate their own settlement. The mediator does not control the negotiations; they merely assist the parties to agree. Sometimes the mediator will give the parties ideas to work from but, generally, only where all the parties agree that the mediator expresses his/her opinion, they may do so. The mediation usually consists of three main tasks: First of all, the mediator needs to settle the conflict. Mediation works on the basis of there being structure in the talks; there is also a timetable and behaviour that 'ordinary' negotiation lacks. The Mediation process is private and confidential and has to be quick. It must promote the amicable agreement on litigations where a solution could not be reached when the customer contacted the customer services of the energy company. So, the presence of a mediator is the key difference in the process. The mediator gives recommendations to the parties, but these are not binding. These characteristics are also the principles of mediation to be respected. Fairness, confidentiality, transparency are essential for the mediator to process.

The mediator will not impose a decision on the parties. Both parties must voluntarily agree to the outcome, which is written in an Agreement, signed and witnessed, which then becomes legally binding on both parties. Second, the mediator often helps to improve the legislation in the sector. As very often mediators face similar problems in many cases, they are in a position to propose reforms of legislation, practices or procedures to facilitate mediation procedures. Third the mediator needs to be in a position to communicate the results of their work. To do so, yearly, they are legally bound to publish an annual report with a catalogue of the number and type of complaints received and their outcome. Furthermore, the mediator must provide consumer associations with regular presentations so that consumers know how to reach them when needed. Another characteristic of mediation is that the parties choose this method to solve their difference on their own will, and the mediator is a person of their choice. Mediation is faster and cheaper than arbitration,[28] and

---

28 See Timothy Martin, "International mediation: an evolving market", *Contemporary Issues in International Arbitration and Mediation: The Fordham*

has a high success rate of settlement.[29] Statistics show that the cost of solving a dispute through mediation process costs less than litigation as well as arbitration. Even though the advantages of mediation are many, still in international disputes, it is not the first choice.[30] This happens for a number of reasons, among which is the lack of familiarity with the process, cultural differences and sometimes large distances.[31] As already mentioned, the mediation process is not a legally binding process; this means that the results of mediation only become binding with a signed written agreement. In order to promote mediation as an ADR tool for international oil & gas disputes, we have to point out the advantages of low cost and quick results and support mediation organizations.

6 Regulators' special competences to resolve disputes

The Third Energy Package contained a provision obliging member states to set up an independent out-of-court dispute resolution scheme for energy complaints. The European Parliament legislative resolution of March 26, 2019, on the proposal for a directive of the European Parliament and of the Council on common rules for the internal market in electricity (recast) [COM(2016)0864 – C8-0495/2016 – 2016/0380(COD)] requires the Member States to ensure that final customers have access to an independent mechanism, such as an energy ombudsman, a consumer body, or a national regulatory authority in order to settle disputes concerning the rights and obligations established under the Directive (Article 26 of the Market Design Directive). The Directive extends the remit of dispute-resolution entities towards all customers, including companies and SMEs. This provision improves consumer confidence by providing access to efficient tools for resolving disputes and obtaining compensation. With the entry into force of the Third Energy Package, the Member States have an obligation to "ensure that there is an independent mechanism, such as an energy ombudsman or consumer body, to deal efficiently with complaints and facilitate out-of-court dispute settlements". Furthermore, the Directive 2013/11/EU,

---

    Papers, ed. Arthur W. Rovine (Leiden-Boston: Martinus Nijhoff Publishers, 2012), pp. 404–417.
29 See Vassilis Hatzopoulos and Sofia Roma, op. cit.
30 Centre for Effective Dispute Resolution (CEDR), "The Fourth Mediation Audit: A Survey of Commercial Mediator Attitudes and Experience" (London: 2010).
31 "The inside Track-how Blue Chips use ADR", *Herbert Smith Freehills*, cit., aggregate number of ADR processes undertaken by 21 surveyed multinational companies in 12-month period.

requires ADR entities, the wider group to which ombudsman schemes belong, to be available for all business-to-consumer disputes.

The Directive allows for greater visibility of the functioning of the free energy market, and so for better regulatory action and business practice. It is very important that an easily identified body is responsible for handling household customers' complaints and dispute settlement, which is usually a third party, proposing or imposing a solution. According to a recent survey of the Association of the Mediterranean Energy Regulators,[32] the majority of MEDREG respondents (80 %) responded positively to the question whether the national regulatory authority was charged with the responsibility to deal with and resolve the household customers' complaints. In the case of Greece, the regulatory authority for energy (RAE) investigates the consumer complaints only related to "regulative supervision issues" mentioned in the Law 4001/2011. Likewise, the Spanish regulatory authority (CNMC) does not hold any competence with regard to disputes emanating from, or initiated by, an individual consumer even though it has competence over disputes taking place between energy market agents (ex. suppliers against TSOs or DSOs). As a result, out of ten respondent countries, in eight countries, the national regulatory authorities are competent to deal with, and resolve, the disputes raised by, or affecting, the household consumers. In one country the national regulatory authority is partially competent, and in another one such authority is not competent. All regulatory authorities in the countries of the Energy Community Contracting Parties have competences to resolve household customers' disputes or complaints. These competences are clearly defined in their primary legislation related to energy and/or consumer protection. In general, there is a high possibility that disputes arise. So regulators are responsible not only for resolving a problem after its creation, by imposing a fine for example, but also for its resolution before it leads to a permanent situation which distorts competition or causes other problems (ex-ante regulation). When asked to resolve a dispute, the regulator has to react quickly, so that welfare loss among stakeholders is minimized. The truth is that while the idea of having an independent regulatory authority for an open to competition market had – among others – the goal

---

32 "Complaint Handling and Dispute Settlement Procedures Available to Household Customers-Joint ECRB-MEDREG Status Review of Applied Practices", *MedReg Working Group on Consumers Issues ECRB Customers and Retail Markets Working Group*, December 2018, http://www.medreg-regulators.org/Portals/_default/Skede/Allegati/Skeda4506-320-2018.12.12/Complaint%20handling_EN.pdf?IDUNI=2jkdndmpum5y5yjq5a2rbsxq9325.

to make sure that governments do not interfere with the liberalized market's issues, sometimes non-independent relationships impede the effectiveness of regulators. This lack of effectiveness has an impact on the consumers' view of regulators as well as of dispute resolution methods. In order to build confidence among consumers so that they will actively participate in the internal energy market, the Third Energy Package considered it of vital importance that the concerns and complaints of consumers are dealt with in a transparent, effective and non-discriminatory manner. To this end, the Member States need to safeguard the existence of an independent mechanism, such as an energy Ombudsman or consumer body, to deal efficiently with complaints and facilitate out-of-court dispute settlements.[33] Consumers must benefit from transparent, simple and inexpensive procedures for dealing with their complaints. This should include a good standard of complaint handling by their energy service providers. Out-of-court dispute settlements should be completed within three months. The Member States must ensure that suppliers effectively communicate to consumers their rights, including information on alternative dispute settlement procedures. Member States should have regard to best practices in complaint handling, in particular in relation to those systems that are available free of charge.

To avoid consumers becoming confused when dealing with the various agents involved in the supply of energy, Member States must ensure that there are single points of contact to provide consumers with all necessary information on their rights and how they can have access to the relevant dispute settlement procedure.[34] Most of the Member States have created departments of dispute settlement and complaints dealing in the functions of the National Energy Regulatory Authorities. A recent example is the French "Independent Administrative Authority" for mediation ("Mediateur National de l'Energie").[35] This new authority for mediation is financed by a part of the contribution of all consumers to public service electricity. According to the French Code of Energy, the mediation process relating to disputes under the jurisdiction of the National Energy Mediator for consumer protection, tariffs, and supply and

---

33 Electricity Directive 2009/72/EC, OJ L 211, 14 August 2009, 55–93, Article 3 sec. 13, Gas Directive 2009/73/EC, OJ L 211, 14 August 2009, 94–136, Article 3 sec. 9.
34 Electricity Directive 2009/72/EC, OJ L 211, 14 August 2009, 55–93, Article 3 sec. 12, Gas Directive 2009/73/EC, OJ L 211, 14 August 2009, 94–136, Article 3 sec. 9.
35 The "Mediateur National de l'Energie" was created under Law 2006-1537 of 7 December 2006, but its function was only recently set under Law 2017-55 of 20 January 2017.

generally contracts not concluded from professionals,[36] must be exercised in a specific period of two months as from the date of the written complaint made with the energy sector company concerned; any referral before the court is suspended. A Mediator is obliged, when the dispute does not fall within its field of competence, to inform the consumer of the rejection of their request for mediation and indicates to them, if necessary, the administrative authority to which it transmits its referral. The MNE also creates general recommendations for consumer protection.[37] The French MNE is willing to contribute to the improvement of the functioning of energy markets; that's why it tries to rebalance the relationship between customers and companies with new ideas and tools. For example, it developed "Energy – info" sections, that include a website and a call centre to provide simple and comprehensive information to consumers. Another innovation of the MNE is the creation of an information and support centre in order to help consumers understand what is happening in the energy and gas sector, and the creation of a programme that compares energy prices and supply offers. According to the annual report of the French MNE, the

---

36 Article L 122-1 provides that: "The national energy ombudsman is responsible for recommending solutions to disputes between natural or legal persons and companies in the energy sector and for helping to inform energy consumers about their rights. It fulfils its mission of consumer mediation under the conditions provided for in Title I of Book VI of the Consumer Code. However, by way of derogation from 3° of Article L 612-2 of this code, referral to another mediator does not preclude the processing of a consumer dispute by the National Energy Mediator that the subject of this dispute falls within its field of competence. It can only be seized of disputes arising from the execution of contracts concluded by a non-professional consumer or by a professional consumer belonging to the category of microenterprises [. . .] These contracts must have already been the subject of a prior written complaint from the consumer to the supplier or distributor concerned, which did not allow the dispute to be settled within a period fixed by regulatory means. It is entered directly and free of charge by the consumer or his representative. It makes its recommendation within a deadline set by regulation and gives reasons for its response. Its referral suspends the prescription of actions in civil and criminal matters during this period. The companies concerned by the litigation provided for in the first paragraph are required to inform their customers of the existence and the methods of referral to the national energy mediator, in particular in the responses to the complaints they receive".
37 "The National Energy Ombudsman: Activity Report 2018", *The French National Energy Ombudsman*, https://www.energie-mediateur.fr/wp-content/uploads/RA_MNE2018_UK.pdf.

volume of referrals sent for resolution is raising, and the mediator has proven itself as a simple, quick, affordable and efficient solution.[38]

7 Which is the best model for Europe to adopt in oil & gas disputes?

After having examined both the mediation and the competence of the regulator as a dispute resolution body, all the given data lead to the conclusion that it would be better for the completion of the internal energy market and the monitoring of trading in the wholesale energy markets of the Union, to create a new Ombudsman specialized in energy issues. The institution of a European Energy Ombudsman would have all the advantages of the function of a mediator, specialization, knowledge of the energy sector, low cost, and would also help to the better protection of the consumers in all energy fields. Mediation is ideally suited for the resolution of many oil & gas disputes for a number of reasons. In the oil & gas sector, as the contracts are of multiannual duration, usually the relationship continues after the resolution of the dispute. As mediation is more friendly than litigation, it offers the parties an opportunity to conclude a settlement on a mutually acceptable basis. Furthermore, in mediation, the parties do not put the blame on each other for what caused the dispute, but through discussion, they are mutually trying to bridge their differences with a mutually accepted binding agreement. The parties in mediation search for common ground. In that spirit, they often find creative solutions that best suit their interests, which, especially in the oil & gas sector, are financially strong. Privacy and confidentiality are essential factors that lead to the success of mediation and make it a clear choice for disputes that the parties would rather not have publicly followed.[39] In cases when the law does not provide for clear solutions for the dispute, the parties would prefer not to go to Court in order to resolve their difference, because a court judgment would create a precedent, and this precedent might be harmful to their interests in the future. So parties opt for mediation instead of litigation because it provides the security of quiet, and 'tailor-made' negotiations.[40] Private procedures in mediation form a twofold advantage; thus

---

38 Ibid.
39 Theodore Panagos, "Handbook of Energy Law" (Athens, Thessaloniki: Sakkoulas Publications, 2015).
40 Lisa C. McManus, "Mediation in Oil and Gas Law Disputes", *LexisNexis Legal Newsroom Staff*, 12 September 2013, https://www.lexisnexis.com/legalnewsroom/energy/b/oil-gas-energy/posts/mediation-in-oil-and-gas-law-disputes.

the outsiders do not get access to any sensitive information, while the parties are protected from the risk of any damaging publicity that may arise out of the proceedings.[41] Furthermore, in the procedures before the Ombudsman, there is the freedom of evidence which provides the parties with the right to present their arguments without the formal restrictions that are set out in the Civil Procedure Code.[42] Also, mediation provides for time and cost efficiency, given the fact that it is generally much quicker and cheaper than taking a case through the Courts. It becomes obvious that this high confidentiality that the nature of the energy sector needs, due to the great volume of business data that are presented during the hearings and the parties do not want to make public, is provided by the ADRs.[43] In conclusion, we would agree that referring to the European Energy Ombudsman offers many advantages such as flexibility in the procedure,[44] given the fact that most of the parties are able to describe in detail this procedure in their ADR clauses or even at the time when the dispute arises. Furthermore, the person(s) deciding the case have expert knowledge and are able to form their conclusions in line with accepted practices. Also, those impartial and high qualified persons can be selected by the parties due to their expertise. Finally, those specialists are fully focused on this particular dispute that has arisen where in litigation a judge may rule upon numerous cases per day. So, parties can be sure that their case will be handled accordingly. In order to benefit from the privilege of international enforceability, mediated settlement agreements must be transformed into arbitral awards. In this regard, it could be said that mediation still depends on arbitration. Yet, if mediated settlement agreements were equally enforceable as arbitral awards, mediation would become a truly independent method of dispute resolution.[45]

---

41 Anthony Conrad K. Kakooza, "Arbitration, Conciliation and Mediation in Uganda: A focus on the practical Aspects", *Uganda Living Law Journal* (2009), Vol. 7, Issue 2, 11–12, https://ssrn.com/abstract=1715664.
42 See Theodore Panagos, op. cit.
43 Ibid., p. 67.
44 Victoria Pernt and Sebastian Lukic, "Mediation on the Heels of Arbitration – Competition or Peaceful Coexistence?", *Kluwer Arbitration Blog*, 6 October 2016, http://arbitrationblog.kluwerarbitration.com/2016/10/06/mediation-on-the-heels-of-arbitration-competition-or-peaceful-coexistence/.
45 See also UNCITRAL, the VIAC, and the International Mediation Institute, held a Joint Conference on 21 September 2016.

Dr. Kyriakos P. Papanikolaou*

16 Energy investments under changed circumstances: Emerging gaps and necessities in arbitration

Abstract: Critical changes of circumstances take place often in long-term energy contracts, for example, contracts on exploration and exploitation of hydrocarbons. If such changes affect the economic equilibrium of a contract, the principle *rebus sic stantibus* may be employed. If the performance of a contractual obligation becomes onerous for unforeseen reasons not connected with the conduct of the contractual parties, a burdened party may seek an adjustment of its obligations or, in extreme cases, even the termination of the contract. *Rebus sic stantibus* is at odds with the principle *pacta sunt servanda* and the sanctity of contracts, which would lead to considerations on change of circumstances in terms of risk distribution between the parties. Even exceptionally within this normative tension, *rebus sic stantibus* might be invoked by the parties in an arbitration process, either as part of a claim or defensively. However, there appears to be some reluctance of parties and tribunals to invoke and apply the principle in arbitration proceedings, including investment arbitration. From the point of view of arbitral tribunals, such reluctance might result from special concerns on their jurisdiction to revise a contract, which relates to considering such claims or objections as judicial disputes or not.

Keywords: Investment arbitration, rebus sic stantibus, force majeure, sanctity of contracts, pacta sunt servanda, risk allocation, renegotiation/adaptation/termination of contracts

1 Long-term energy contracts under a change of circumstances

Energy investments are usually based on long-term contracts. The most crucial aspect of such contracts is time, which in turn means a possible change of circumstances. The parties conclude a contract taking into account the current circumstances as a necessary context of their contractual will. At best, they might draft the terms of the contract, e.g. on price rights and obligations, foreseeing some change of the context. However, it is rather impossible that parties

---

* Lecturer of Public Law at the Law School of Democritus University of Thrace – Attorney at the Greek Supreme Courts.

of energy contracts lasting for many years, usually decades, could foresee, even briefly, all the substantial changes that would have affected their contractual will.[1] For instance, a contract on exploration and exploitation of hydrocarbons provides for price rights between the parties based on a normal, in any case foreseeable, fluctuation of the relevant prices in the world market. However, the margins of such fluctuation are not guaranteed. So, a radical change of circumstances may affect the economic equilibrium established between the parties through the energy contract. The obvious question arising is whether there might be any legal consequences on the performance of contractual obligations in such cases of changed circumstances. If an affirmative answer cannot be excluded, then the central question is what kind of change and what level of change is necessary for legal consequences to emerge; and, following that, what are the legal consequences and how they come into play; are they automatic, are they depended upon the existence of explicit contractual terms, and – probably the most important for the present discussion – what is the role of arbitral tribunals in cases involving such claims based upon the changed circumstances? All these questions are essential for arbitration since this is normally the mechanism for the resolution of such energy disputes. Moreover, vice versa, probably the fact that such disputes are subject to review by arbitral tribunals – and not by judges – is a key parameter for the mechanism of triggering the legal consequences. For instance, recent contracts on oil exploration and exploitation in Greece provide for a settlement of the relevant disputes by arbitration conducted in accordance with the Rules of the I.C.C.[2]

2 The distinction between *rebus sic stantibus* and *force majeure*

A well-known distinction should be reminded at this point: a change of circumstances does not equate with *force majeure*. *Force majeure* makes it impossible for one or more parties to perform their contractual obligations. On the other hand, a change of circumstances makes that performance – not impossible, but – onerous, or extremely onerous. It is not an act of God, a state of war, or a revolution. Such extreme and objective situations are something more. On the other hand, the change of circumstances is distinguished

---

1   See P. Bernardini, "Stabilization and Adaptation in Oil and Gas Investments", *Journal of World Energy Law and Business* 1, no. 1 (2008), 98 et seq., at 98.
2   See Article 23 §§ 3–11 of the Contracts on Petroleum Operations in the contract areas Ionian (Western Greece), Sea Area 10 (Ionian Sea) and Sea Area West of Crete, as they have been ratified by Statutes No. 4629/2019, 4630/2019 and 4631/2019, respectively.

from the discovery by a party that the conditions on which it has concluded a contract were false. Such a subjective discovery of something already existent is less than a change of circumstances, with the latter being usually codified as the principle *rebus sic stantibus*. The parties may have included a hardship clause in the contract. Usually, they do not, whereas they agree on explicit *force majeure* clauses. For example, in the recent Greek contracts of oil exploration and exploitation, a *force majeure* clause is explicitly provided,[3] whereas no general hardship clause is included. Should we consider such silence as a purposeful choice or is it a lacuna that should be aptly filled? Adversely, might a *rebus sic stantibus* clause be considered as always implied? Moreover, even in cases where a hardship clause is provided, should we consider it as an indicative illustration of a general principle applied in any event, regardless of it being explicitly provided or not? The normative impact of such general questions is intensified by the very fact that *rebus sic stantibus* is not a rule provided in all legal systems;[4] therefore, it is not sure that, even without an explicit contractual term, this principle shall be applied through the general law governing the contract.[5]

3 *Rebus sic stantibus* vs. *Pacta sunt servanda*

The *pacta sunt servanda* principle is at odds with the application of the maxim *rebus sic stantibus*. This contradiction is more precise in an arbitration context since arbitration by its very essence is an institution established by the contractual will of the parties. In principle, not only the arbitration clause but the whole contract is to be served, indifferently to any change of the initial circumstances.

---

3 Cf. Article 26 of the Contracts on Petroleum Operations referred to *supra* note 2.
4 For a comparison between common law and civil law orders on the issue see A. Kolo and T. W. W. Wälde, "Renegotiation and Contract Adaptation in International Investment Projects. Applicable Legal Principles and Industry Practices", *The Journal of World Investment* 1, no. 1 (July 2000), 5 et seq., at 34–38.
5 Cf. G. Kaufmann-Kohler, "Arbitral Precedent: Dream, Necessity of Excuse? (The 2006 Freshfields Lecture)", *Arbitration International* 23, no. 3 (2007), 357 et seq., at 363–364, on *arbitral jurisprudence*, consisting of general principles deriving from national law and international treaties, and not from arbitral case law *per se*. For this reason, domestic law keeps its significance for the dynamics of *lex mercatoria*, wherein incarnations of basic domestic rules and principles have been filtered; on this process of normative transplantation see J. H. Dalhuisen, "Legal Orders and Their Manifestation: The Operation of the International Commercial and Financial Legal Order and Its Lex Mercatoria", *Berkeley Journal of International Law* 24, no. 1 (2006), 129 et seq., at 184–187.

The sanctity of the contract is, in principle, indifferent to facts disturbing the contractual balance between the parties. This status might be considered as not being affected even in cases of long-term contracts. It could be argued that even long-term contracts incorporate, even implicitly, a distribution of risk between the parties.[6] That risk is a burden allocated to each party, which refers *inter alia* to the possible changes in the contractual context. If such a change occurs, the party that is negatively affected should not claim that it no longer bears the obligations provided by the contract. On the contrary, the change should be considered as a fulfilment of the risk allocated to that party. As far as the risk becomes a reality, the party bearing the relevant burden has to fulfil its contractual obligations under the changed circumstances, since that burden is equally a contractual term, even if it is only implied as part of the *pacta sunt servanda* principle. For the same reasons, even if hardship clauses are included in a contract, they should be interpreted in a strict manner, without being considered as express manifestations of a broader *rebus sic stantibus* principle applied in the case.

In any case, even if *rebus sic stantibus* is applied only under strict conditions, because it is heavily counter-balanced by *pacta sunt servanda*, still it does remain a legal principle and, hence, the possibility of its *exceptional* application is retained.[7] Such exceptional use of the principle is reflected not only in arbitral awards but also explicitly in basic international law instruments. For example, Article 62 of the Vienna Convention on the Law of Treaties provides that a fundamental change of circumstances may be invoked as ground for terminating a treaty or withdrawing from a treaty under the following cumulative conditions:[8] (a) the change was not foreseen by the parties; (b) the circumstances existed at

---

6 Cf. J. v. Dunné, "The Changing of the Guard. *Force majeure* and Frustration in Construction Contracts: The Foreseeability Requirement Replaced by Normative Risk Allocation", *International Construction Law Review* 19, no. 2 (2002), pp. 162 et seq., passim, where it is maintained that the foreseeability requirement of *force majeure* and frustration is being gradually replaced by risk allocation in the construction sector.

7 See R. Kos and M. Durbas, "The Arbitrators' (Perceived) Power to Revise a Contract vs. the Power of the Public Policy Clause", *Arbitration International* 17 (2001), pp. 135 et seq., at 141, where *rebus sic stantibus* is regarded as a narrowly interpreted and exceptionally applied principle in the light of the cornerstone *pacta sunt servanda*.

8 See D. J. Bederman, "The 1871 London Declaration, Rebus Sic Stantibus and a Primitivist View of the Law of Nations", *The American Journal of International Law* 82, no. 1 (January 1988), 1 *et seq.*, at 38, where it is pointed out that under Article 62

the time of the conclusion of the treaty constituted an essential basis of the consent of the parties to be bound by the treaty; (c) the effect of the change is radically to transform the extent of obligations still to be performed under the treaty; (d) the fundamental change is not the result of a breach by the party invoking it, either of an obligation under the treaty or of any other obligation owed to any other party. The above four conditions are generally accepted,[9] as necessary for any application of the principle *rebus sic stantibus*.[10] The principle may be invoked by the parties within an arbitration procedure either as part of a claim or defensively. A party may seek termination or an adjustment of its contractual obligations, if the economic equilibrium of the contract has been substantially modified for reasons beyond the control of the parties. More interesting, especially for investment arbitration, is the reverse situation, namely the case of a State that has already modified a contract with an investor in the energy sector for reasons reflecting the principle *rebus sic stantibus*. In such cases, the tribunal has to decide on defensive uses of the *rebus sic stantibus* principle.[11] However, the parties are usually reluctant to rely upon arguments derived from the *rebus sic stantibus* principle, even if they use its vocabulary.

4 The reluctance to invoke the principle *rebus sic stantibus* in arbitration procedures

As an illustration of this stance on the principle, a relevant ICSID arbitration case could be assessed, the *Burlington Resources v. Republic of Ecuador*,[12] which illuminates the sensitivity of the issue and the reluctance of the parties

---

of the Vienna Convention the *rebus sic stantibus* doctrine allows states to alter of cancel treaty obligations "within substantial definitional limits".

9 There is a fifth condition for the application of Article 62 of the Vienna Convention, which is not mentioned above as it cannot be applied but in relation to international treaties: A party shall not terminate or withdraw from a treaty establishing boundaries.

10 For a general approach to *rebus sic stantibus* as a principle of international treaties see O. J. Lissitzyn, "Treaties and Changed Circumstances (Rebus Sic Stantibus)", *The American Journal of International Law* 61, no. 4 (October 1967), 895 et seq., passim.

11 Infra at IV.

12 ICSID Case No. ARB/08/5, *Burlington Resources Inc. (The Claimant) v. Republic of Ecuador (The Respondent)*, Decision on Liability, 14 December 2012 – Arbitral Tribunal: G. Kaufmann-Kohler (Pr.), Br. Stern (Arb.), Fr. Orrego Vicuña (Arb.).

to invoke the principle.[13] In the 1980s, *Ecuador* had adopted the so-called service contract model for the exploitation of hydrocarbons, where the government remained the sole owner of any oil produced in the exploration Block. The contractor had the right only to reimbursement of its costs and a fee. In 1992 a new model was introduced, the 'production-sharing contract' (PSC). The contractor assumed the entire risk of the exploration and exploitation and would, in exchange, receive a share of the oil produced. That model would put an end to the excessive and inefficient costs and would attract foreign investment. In 2001 Burlington Resources acquired interests in exploration and exploitation in two Blocks. It also formed a tax consortium with another company (Perenco), which had interests in the same Blocks, to share the relevant taxes. The PSCs contained participation formulas allocating the oil produced between Ecuador and the contractors based on volumes of oil produced. In the arbitration, Burlington Resources claimed that the participation formula was grounded solely on the volume and quality of oil produced. Ecuador, on the other hand, submitted that the participation formulas were also premised on the price of oil at the time of the PSCs.[14] Ecuadorian law was applicable, according to the contracts and Article 42 (1) of the ICSID Convention.[15] The PSCs incorporated tax clauses regulating the tax treatment of the contractor, who was obliged to pay, among others, an income tax of 25 %, which was under tax modification clauses calling for the application of correction factors depending on tax increases or decreases with an impact on the economy of the contract.[16] In 2002 oil prices began to increase and this rise was maintained until 2008.[17] According to Ecuador, the increase in oil prices was unforeseen and unforeseeable, destroyed the economic stability of the contracts and that was no longer fair, since the State as the owner of the oil should benefit from the increase in oil prices to a greater extent than the contractor. That benefit was not possible since the PSCs allocated the majority of oil production to the contractor.[18] After a failure of renegotiation, a new bill (Law 42) was enacted in 2006, which granted the Ecuadorian

---

13 Cf. R. Kos/M. Durbas, "The Arbitrators' (Perceived) Power", supra note 7, at 146, on the reluctance of arbitral tribunals to revise a contract, even if they have been granted with the relevant power.
14 See *Burlington Resources*, supra note 12, at § 17.
15 The domestic law was applicable complementarily to the first and foremost applicable US-Ecuador BIT. Ibid., at §§ 178–179.
16 Ibid., at § 21.
17 Ibid., at § 24.
18 Ibid., at §§ 26–27.

State participation of at least 50 % over the extraordinary revenues (unforeseen surpluses) caused by increases on the market price of oil. This additional 50 % was a tax, which was considered by the Constitutional Court as constitutional. In November 2006, the Law 42 tax rate was increased from 50 % to 99 %.[19] In 2007 a renegotiation process was announced, which finally failed. In 2008 Burlington Resources filed a Request for ICSID arbitration against Ecuador.[20]

From the above, it appeared that Ecuador's arguments would rely on the principle *rebus sic stantibus*. Nevertheless, there was a disagreement between the parties on whether Ecuador relied on the principle. Ecuador denied that its arguments were manifestations of the principle, whereas Burlington Resources argued for the opposite.[21] The arbitral tribunal recognized the principle and referred to its three elements: (a) Imbalance because of extraordinary and unforeseeable or unforeseen events, (b) Severity of the imbalance rendering the performance of the contract too burdensome, (c) The extraordinary event is not a consequence of actions or omissions of the party invoking the principle.[22] Moreover, the arbitral tribunal noted that there were certain references to the principle or its elements in certain documents on record. Nevertheless, the Court finally accepted that there was no need to examine the facts under the *rebus sic stantibus* perspective, because Ecuador in its post-hearing brief expressly disclaimed reliance on that principle and, instead, presented its acts and the dispute as a matter of contract interpretation concerning economic premises integrated into the contract.[23] The power of *pacta sunt servanda* precluded – not the tribunal, but – the State party from making the dispute a matter of *rebus sic stantibus*.

As can be expected, the reluctance to rely upon the principle would be enhanced in cases where the *rebus sic stantibus* arguments are put forward not defensively but as a claim. What are the consequences of a successful reliance on the principle? Either the contract shall be suspended, or it shall be terminated in extremely severe cases. In cases where the contract provides for a renegotiation initiative under certain circumstances, the parties may come to an adaptation of the contractual obligations to the changed circumstances. If such a renegotiation unreasonably fails and the arbitral tribunal comes to the conclusion

---

19 Ibid., at §§ 30–32, 34–35.
20 Ibid., at §§ 50, 67.
21 Ibid., at §§ 103, 105, 358, 380, 406.
22 Ibid., at § 406 note 659.
23 Ibid., at §§ 407–408.

that *bona fide* it cannot be reasonably expected that the party can perform its contractual obligations without a proper adaptation to the new circumstances, then the arbitral tribunal may terminate the contract. So, *rebus sic stantibus* is the *ultimum refugium* of the parties in long-term contracts. A final issue relates to the consequences of arbitral awards which do not only terminate or suspend a contract but adapt the contractual terms to the changed circumstances to restore the contractual equilibrium. For such an adaptation, in other words for an arbitral modification of the contract, the arbitrators should rely either on the arbitration clause or on the *lex arbitri* and the *lex contractus*. However, even if the arbitral tribunal accepts such an arbitral power, the adjudicative nature of such awards is at least contested.[24] Strong doubts about the nature of arbitral revisions of contractual terms are reflected on the questionable enforcement of such awards, at least with the direct and binding manner secured by the leading international conventions (e.g. the New York Convention).[25] For the same reason, the jurisdiction of ICSID arbitrators may be questionable, under Article 25 of the Convention, in cases about claims to an arbitral revision of contractual terms.[26]

---

24 See K. P. Berger, "Renegotiation and Adaptation of International Investment Contracts: The Role of Contract Drafters and Arbitrators", *Vanderbilt Journal of Transnational Law* 36, no. 4 (October 2003), 1347 et seq., at 1371–1373, where a distinction is made between cases on the adjustment of contractual terms and cases on the interpretation of renegotiation clauses.
25 See P. Bernardini, "Stabilization and Adaptation", supra note 1, at 108.
26 Article 25 (1) of ICSID reads as follows: "The jurisdiction of the Centre shall extend to any *legal dispute* arising directly out of an investment [. . .]". On this aspect of the interpretation of Article 25 (1) ICSID see K. P. Berger, "Renegotiation and Adaptation", supra note 24, at 1374–1375.

Veröffentlichungen des Instituts für deutsches und europäisches Wirtschafts-, Wettbewerbs- und Regulierungsrecht der Freien Universität Berlin, ab Band 48 fortgeführt vom **Institut für Energie- und Regulierungsrecht Berlin**

Herausgegeben von Franz Jürgen Säcker:

Band 1		Franz Jürgen Säcker (Hrsg.): Deutsch-russisches Energie- und Bergrecht im Vergleich. Ergebnisse einer Arbeitstagung vom 31. März / 1. April 2006. 2007.
Band 2		Franz Jürgen Säcker / Walther Busse von Colbe (Hrsg.): Wettbewerbsfördernde Anreizregulierung. Zum Anreizregulierungsbericht der Bundesnetzagentur vom 30. Juni 2006. 2007.
Band 3		Dirk Zschenderlein: Die Gleichbehandlung der Aktionäre bei der Auskunftserteilung in der Aktiengesellschaft. Zum Problem der Zulässigkeit der Weitergabe von Informationen an einzelne Aktionäre und Dritte. 2007.
Band 4		Simone Kirchhain: Die Anwendung der Vertikal-GVO auf innerstaatliche Wettbewerbsbeschränkungen nach der 7. GWB-Novelle. 2007.
Band 5		Franz Jürgen Säcker: Der Independent System Operator. Ein neues institutionelles Design für Netzbetreiber? 2007.
Band 6		Stefanie Otto: Allgemeininteressen im neuen UWG. § 1 S. 2 UWG und die wettbewerbsfunktionale Auslegung. 2007.
Band 7		Jochen Eichler: Vertragliche Dritthaftung. Eine Auseinandersetzung mit der Frage der Dritthaftung von sogenannten Experten und anderen Auskunftspersonen im Rahmen des § 311 Abs. 3 BGB. 2007.
Band 8		Markela Stamati: Die Anforderungen der operationellen Entflechtung nach den Beschleunigungsrichtlinien der Europäischen Kommission. Umsetzung in Deutschland und Griechenland. 2008.
Band 9		Franz Jürgen Säcker: The Concept of the Relevant Product Market. Between Demand-Side Substitutability and Supply-Side -Substitutability in Competition Law. 2008.
Band 10		Renate Rabensdorf: Die Durchgriffshaftung im deutschen und russischen Recht der Kapitalgesellschaften. Eine rechtsvergleichende Untersuchung. 2009.
Band 11		Franz Jürgen Säcker: Der beschleunigte Ausbau der Höchstspannungsnetze als Rechtsproblem. Erläutert am Beispiel der 380-kV-Höchstspannungsleitung Lauchstädt – Redwitz – Grafenrheinfeld mit Querung des Rennsteigs im Naturpark Thüringer Wald. 2009.
Band 12		Helen Mahne: Eigentum an Versorgungsleitungen. 2009.
Band 13		Franz Jürgen Säcker (Hrsg.): Russisches Energierecht - Gesetzessammlung. 2009.
Band 14		Franz Jürgen Säcker / Maik Wolf: Integrierte Energieversorgung in geschlossenen Verteilernetzen. Zum Gestaltungsspielraum des Gesetzgebers zur Neuregelung des § 110 EnWG im Lichte des Dritten EG-Energiepakets. 2009.
Band 15		Franz Jürgen Säcker (Hrsg.): Das Dritte Energiepaket für den Gasbereich. Deutsch-Englische Textausgabe mit einer Einführung. 2009.
Band 16		Franz Jürgen Säcker (Hrsg.): Das Dritte Energiepaket für den Elektrizitätsbereich. Deutsch-Englische Textausgabe mit einer Einführung. 2009.

Band 17 Thomas Dörmer: Die Unternehmenspacht. Rechtsstellung der Vertragsparteien unter besonderer Berücksichtigung der Pflicht des Unternehmenspächters zur ordnungsgemäßen Unternehmensführung sowie der Rechtslage bei Vertragsbeendigung. 2010.

Band 18 Klaas Bosch: Die Kontrolldichte der gerichtlichen Überprüfung von Marktregulierungsentscheidungen der Bundesnetzagentur nach dem Telekommunikationsgesetz. 2010.

Band 19 Geng-Sook Leem: Einheitliche Corporate Governance-Grundsätze für die Europäische Aktiengesellschaft (SE). Eine rechtsvergleichende Untersuchung anhand der Ausgestaltung der SE im deutschen und britischen Recht. 2010.

Band 20 Wiebke Gebhardt: Gentechnik und Koexistenz nach der Gesetzesnovelle von 2008: Zivilrechtliche Haftung im Vergleich Deutschland und USA. 2010.

Band 21 Cathrin Isenberg: Die Geruchsmarke als Gemeinschaftsmarke. Schutzfähigkeit und Einsatzmöglichkeiten. 2010.

Band 22 Franz Jürgen Säcker / Jochen Mohr / Maik Wolf: Konzessionsverträge im System des europäischen und deutschen Wettbewerbsrechts. 2011.

Band 23 Judith Antonia Loeck: Die unzumutbare Belästigung nach der UWG Novelle 2008 und dem Gesetz zur Bekämpfung unerlaubter Telefonwerbung und zur Verbesserung des Verbraucherschutzes bei besonderen Betriebsformen. 2011.

Band 24 Jörg Jaecks: Konzernverrechnungsklauseln und verwandte einseitige Verrechnungsbefugnisse im Mehrpersonenverhältnis. 2011.

Band 25 Franz Jürgen Säcker: Marktabgrenzung, Marktbeherrschung, Markttransparenz und Machtmissbrauch auf den Großhandelsmärkten für Elektrizität. 2011.

Band 26 Susanne Wende: Die einheitliche Auslegung von Beihilfen- und Vergaberecht als Teilgebiete des europäischen Wettbewerbsrechts. 2011.

Band 27 Leonie Kempel: Die Anwendung von Art. 102 AEUV auf geistiges Eigentum und Sacheigentum. Die Voraussetzungen des Kontrahierungszwangs nach Art. 102 AEUV und der Essential-Facility-Doktrin unter besonderer Berücksichtigung der Unterschiede zwischen geistigem Eigentum und Sacheigentum. 2011.

Band 28 Christoph Schuldt: Werbejingles – schützenswerte Kompositionen!? Die urheberrechtliche und markenrechtliche Schutzfähigkeit von Werbejingles vor unbefugter Nachahmung. 2012.

Band 29 Lydia Scholz: Die Rechtfertigung von diskriminierenden umweltpolitischen Steuerungsinstrumenten. Eine Untersuchung der Reichweite der Warenverkehrsfreiheit und ihrer Begrenzung durch den Umweltschutz als Vertragsziel am Beispiel der deutschen Energieförderungsgesetze EEG und KWKModG. 2012.

Band 30 Franz Jürgen Säcker: Investitionen in Kraftwerke zur Sicherung einer zuverlässigen Elektrizitätsversorgung nach der Energiewende. Rechtliche und ökonomische Rahmenbedingungen. 2012.

Band 31 Florian Leib: Kartellrechtliche Durchsetzungsstrategien in der Europäischen Union, den USA und Deutschland. Eine rechtsvergleichende Darstellung. 2012.

Band 32 Holger Hoch: Marktverschlusseffekte und Effizienzen vertikaler Zusammenschlüsse. Kartellrechtliche Beurteilung nach europäischem und deutschem Recht. 2012.

Band 33 Sebastian Kemper: Gasnetzzugang in Deutschland und in Spanien. 2012.

Band 34 Elena Timofeeva: Unbundling in der russischen Elektrizitätswirtschaft im Vergleich zum deutschen und europäischen Energierecht. 2012.

Band	35	Gisela Drozella / Harald Krebs: Marktbeherrschung im Bereich Stromerzeugung / Stromgroßhandel. Eine kritische Analyse der neueren Sicht des Bundeskartellamts. 2013.
Band	36	Christian Rehm: Die Einzel- und Gesamtverantwortung der Vorstandsmitglieder der Aktiengesellschaft. Die Verantwortung für die Leitung und Geschäftsführung im mehrköpfigen Vorstand in der unabhängigen und der herrschenden AG. 2013.
Band	37	Ann-Christin Richter: Mengen- und umsatzbezogene Rabatte marktbeherrschender Unternehmen in den Grenzen des Art. 102 AEUV. 2013.
Band	38	Nikolaos Pitsos: Die erhebliche Behinderung des wirksamen Wettbewerbs (SIEC-Test) im Fusionskontrollrecht. Zugleich ein Beitrag zur Problematik der oligopolistischen Marktbeherrschung unter Vergleich mit dem amerikanischen SLC-Test. 2013.
Band	39	Mirko Heinemann: Das Markenzeichen als Anknüpfungspunkt des Markenimages. 2013.
Band	40	Sylvia Lorenz: Der Immaterialgüterschutz virtueller Ladengeschäfte. Ein Vorschlag für die rechtliche Erfassung virtueller Güter. 2014.
Band	41	Anke Reimers: Wettbewerbsrechtliche und regulatorische Analyse der Kapazitätenvergabe an Grenzkuppelstellen im Elektrizitätsbinnenmarkt. 2014.
Band	42	Franz Jürgen Säcker / Carsten König / Lydia Scholz: Der regulierungsrechtliche Rahmen für ein Offshore-Stromnetz in der Nordsee. 2014.
Band	43	David Preisendanz: Der Humor in der vergleichenden Werbung. 2014.
Band	44	Martin Gerig: Vollendung des EU-Energiebinnenmarktes vs. nationale Marktabschottungen. Europarechtskonformität mitgliedstaatlicher Fördermaßnahmen und Kapazitätsmärkte am Beispiel Deutschlands. 2014.
Band	45	Jan Patrick Ehinger: Dritthaftung für Kunstexpertisen und Aufnahmebestätigungen in den Catalogue raisonné. Ein Beitrag zur Expertenhaftung. 2014.
Band	46	Julia Gerzen: Das Recht der Gesellschafter-Fremdkapitalfinanzierung. Rechtsvergleichende Analyse der Behandlung von Gesellschafterdarlehen vor und in der Insolvenz der GmbH nach deutschem und russischem Recht. 2014.
Band	47	Laura Börger: Die gerichtliche Kontrolldichte bei der Überprüfung von Entscheidungen der Europäischen Kommission auf dem Gebiet der Fusionskontrolle. 2014.

Veröffentlichungen des Instituts für Energie- und Regulierungsrecht Berlin

Band 48–65 herausgegeben von Franz Jürgen Säcker

ab Band 66 gemeinsam mit Jochen Mohr

Band	48	Daniel Simon Travers: Kopplungsangebote. Eine Studie zum Vertrags- und Lauterkeitsrecht. 2015.
Band	49	Anna Olbryś-Sobieszuk: EU-rechtliche Schranken der Glücksspielwerbung in Deutschland, Großbritannien und Polen. 2015.
Band	50	Franz Jürgen Säcker/Lydia Scholz/Thea Sveen: Renewable Energy Law in Europe. Challenges and Perspectives. 2015.
Band	51	Lina Berezgova: Festlegung von Stromnetzentgelten in Deutschland und Russland. Eine rechtsvergleichende Untersuchung der anreizbasierten Regulierungsvorgaben. 2015.

Band	52	Fengliang Jin: Aufgreifkriterien in der Fusionskontrolle. Eine rechtsvergleichende Untersuchung des europäischen (Art. 3 FKVO), deutschen (§ 37 GWB), taiwanesischen (§ 6 FHG) und festlandchinesischen (§ 20 AMG) Rechts. 2015.
Band	53	Vasileios Triantafyllidis: Die Einordnung der Passing-On-Defense in das allgemeine Schadensersatzrecht. Eine rechtsvergleichende, dogmatische Studie. 2015.
Band	54	Peipei Liu: Die Inhaltskontrolle von Allgemeinen Geschäftsbedingungen und formularmäßigen Vertragsklauseln im deutsch-chinesischen Vergleich. 2016.
Band	55	Evelyn Kozak: Zur Notwendigkeit eines arbeitsrechtlichen und haftungsrechtlichen Whistleblowerschutzes. 2016.
Band	56	Genevieve Baker: Der bereicherungsrechtliche Nutzungsanspruch. 2017.
Band	57	Maria Zaykova: Der rechtliche Rahmen für den bergbaulichen Zugang zu den Erdöl- und Erdgaslagerstätten in der Russischen Föderation im Vergleich zum deutschen Bergrecht. 2017.
Band	58	Qian Luan: Geldschöpfungsprozess und Geld als abstraktes Schuldversprechen. Die europäisch-verwaltungsrechtliche Betrachtung des Geldschöpfungsprozesses der EZB und die Rechtsnatur von Zentralbankgeld und Geschäftsbankengiralgeld. 2017.
Band	59	Sára Gabriella Hoffman: Regulation of Cloud Services under US and EU Antitrust, Competition and Privacy Laws. 2017.
Band	60	Matthias Schöpf: Das neue Planungsrecht der Übertragungsnetze: Vorgaben des deutschen und europäischen Rechts. 2017.
Band	61	İpek Çevik: Eine vergleichende Untersuchung der Joint Ventures – Gemeinschaftsunternehmen – nach europäischem, türkischem und deutschem Wettbewerbsrecht. Die Analyse von wettbewerbsrechtlichen Auswirkungen bei Gründung der konzentrativen und/oder kooperativen Gemeinschaftsunternehmen. 2017.
Band	62	Raffaele Mazza: Verbotsirrtum und Vertrauen auf Rechtsrat im europäischen Kartellrecht. Eine rechtsvergleichende Analyse des europäischen, des deutschen und des französischen Kartellrechts. 2017.
Band	63	Stavroula Antoniou: Die Kontrolle einseitiger Preisanpassungsrechte in Dauerschuldverhältnissen am Beispiel langfristige Energielieferverträge. Eine rechtsvergleichende Untersuchung nach deutschem und griechischem Recht. 2018.
Band	64	Lucas Noura de Moraes Rêgo Guimarães: Energieversorgungssicherheit im Europarecht mittels der Förderung erneuerbarer Energien und der Interkonnektion der Netze. 2018.
Band	65	Nino Kobadze: Regulation of Access to Gas Networks in Russia in Comparison with the Energy Law of Germany and the EU. 2018.
Band	66	Felix Krone: Das Recht der unkörperlichen Verwertung für das Internet als Teilhaberecht. Zugleich eine Grundsatzschrift für ein kommunikationsoffenes Urheberrecht 2018.
Band	67	Katja Stefanie Middelhoff: Das Spürbarkeitsmerkmal des UWG 2015. 2019.
Band	68	Tatjana Schleicher: Die Umsetzung des § 14a EnWG zwischen Markt und Regulierung im neuen Strommarktdesign. 2020.
Band	69	Eleni-Aristea Alevizou: Der Rechtsrahmen für die nachhaltige Entwicklung der Photovoltaikenergie in Deutschland und in Griechenland. 2020.
Band	70	Rui Long: Die Aufgreifkriterien der chinesischen Fusionskontrolle. Minderheitsbeteiligung, Gemeinschaftsunternehmen, konzerninterne Umstrukturierung und neue Fragen zum digitalen Markt aus rechtsvergleichender Sicht mit deutschem und europäischem Recht. 2020.